THE METAPHYSICS OF
QUANTUM THEORY

THE METAPHYSICS OF QUANTUM THEORY

Henry Krips

CLARENDON PRESS · OXFORD

1987

Oxford University Press, Walton Street, Oxford OX2 6DP

Oxford New York Toronto
Delhi Bombay Calcutta Madras Karachi
Petaling Jaya Singapore Hong Kong Tokyo
Nairobi Dar es Salaam Cape Town
Melbourne Auckland

and associated companies in
Beirut Berlin Ibadan Nicosia

Oxford is a trade mark of Oxford University Press

British Library Cataloguing in Publication Data

Krips, Henry
The metaphysics of quantum theory.
1. Quantum theory   2. Physics—Philosophy
I. Title
530.1′2′01   QC174.13
ISBN 0-19-824971-3

Library of Congress Cataloging in Publication Data

Krips, Henry.
The metaphysics of quantum theory.
Bibliography: p.
Includes index.
1. Quantum theory. 2. Metaphysics. I. Title.
QC174.13.K75 1987   530.1′2   87-12965
ISBN 0-19-824971-3

Phototypeset by Macmillan India Ltd, Bangalore 25
Printed and bound in Great Britain by
Biddles Ltd, Guildford and King's Lynn

For Valerie

Preface

CHAPTERS 1, 2, 3 and 6 are non-technical, and focus on the metaphysical issue of interpreting the theoretical entities of Quantum Theory: the quantum systems themselves, the state-vectors, and the probabilities. Chapters 4 and 5 deal with the classic paradoxes of Quantum Theory—Schrödinger's cat and the Einstein–Podolski–Rosen paradox—as well as the problems surrounding the measurement process more generally. These chapters are a little more technical, and in particular introduce the operator form of representation for physical quantities. Purely technical issues are discussed in the appendices however, with for example an appendix on von Neumann's representation of continuous-valued physical quantities. Chapters 7 to 9 focus on more recent philosophical issues in the foundations of Quantum Theory, issues such as determinism, locality, and the relation between modal and statistical discourse. In order to improve readability of the text, an index of the more extensively used principles is included.

Acknowledgements

In writing this book I have been greatly assisted by the hospitality extended to me every Lent term over the past few years by the Department of History and Philosophy of Science, University of Cambridge, and the Master and Fellows of Darwin College. I am particularly indebted to Jeremy Butterfield and Hugh Mellor of the Cambridge Philosophy Department, for reading and criticizing several earlier drafts of this work. I would also like to thank Michael Redhead, David Papineau, Linda Burns, David Wood, and Martin Leckie for discussions on particular points, Valerie Krips for general editorial work and encouragement, the Committee for Research and Graduate Studies of the University of Melbourne, who generously funded the production of the manuscript, and Rita Hutchison for typing it.

Contents

It gives me great pleasure to read your careful investigations the result of which fully agrees with my own view on this matter: in the concrete individual case one has to ascribe real existence both to the wave field and to the (more or less) localised quantum unless one is ready to admit a telepathic coupling between objects in different regions of space . . .

A. Einstein, 1953

(letter to M. Renninger, 3 May 1953, quoted in Jammer (1974) p. 494.)

Your view is precisely the same as the one I have espoused from the beginning (1926): both particles and waves have some sort of reality, but it must be admitted that the waves are not carriers of energy or momentum . . .

M. Born, 1955

(letter to M. Renninger, 23 May 1955, quoted in Jammer (1974) p. 495.)

Introduction

HISTORICALLY the interpretation of QT (non-relativistic Quantum Theory) has been dominated by two main traditions. There is the 'orthodox interpretation' (or Copenhagen interpretation) which grew out of the views of Bohr and has been promulgated by Heisenberg, Born, Pauli, *et al.* And there is the 'realist interpretation' which developed from Einstein's criticisms of Bohr, and has been forwarded by Landé, Popper, Putnam, Bohm, and more generally the hidden variables school.

The latter group object to the orthodox interpretation on the grounds of its anti-realism, and indeed Bohr was anti-realist in his attitude to the formalism of the new quantum theory, at least in the respect that he took the state-vectors of systems to be purely heuristic devices. Moreover Heisenberg was anti-realist in respect of his endorsement (if only for heuristic purposes) of verificationist principles, particularly in his early writings. But as we shall see neither Heisenberg nor Bohr were anti-realists in the metaphysical sense of denying the existence of an objective external reality lying behind the 'veil of perception', nor did they eschew the scientific realist's commitment to describing that reality within science. In particular they shared with Einstein and the 'realists' a belief in the objective reality of atoms, as well as putting forward atomic theories of matter within science. In short the disagreement between Bohr and Heisenberg on the one hand and Einstein on the other was not a disagreement over metaphysical realism or indeed scientific realism. Rather it was a disagreement about the terms in which external reality was to be described.

Bohr took it that the description of reality had ultimately to be in classical terms, although a degree of imprecision (or indeterminacy) in fitting the classical descriptions to reality was to be allowed; indeed QT was to be the site at which this degree of imprecision was to be theorized. Moreover the statistical nature of QT was taken to be rooted in this degree of imprecision, i.e. it is because we can only say imprecisely what goes on that we can only predict with a probability less than certainty what the results of certain experiments will be.

But for Einstein, as opposed to Bohr, this degree of imprecision in the fit of classical concepts to quantum phenomena only signalled the need to replace the stock of classical descriptions used in Bohr's QT with some new descriptions, which would fit precisely. Moreover Einstein hoped to be able to eliminate the statistical turn which physics had taken at a fundamental level with the advent of QT.

Bohr's followers, Heisenberg, Born, *et al.*, went beyond Bohr however in suggesting new non-classical descriptions for atoms in terms of state-vectors. These new descriptions were unacceptable to Einstein because they cemented the role of probabilities within QT. Indeed Born referred to atoms as guided by 'probability fields'. Within this new QT Bohr's followers retained the notion of indeterminate values for physical quantities. Indeed the new style of description in terms of state-vectors provided a criterion for this indeterminacy (Q is indeterminate in value in system S at time t iff S at t has a state-vector which is not an eigenvector of Q). In short, the new orthodoxy for QT retained Bohr's idea of QT as the site for detailing the limits of classical descriptions, but embedded these classical descriptions within a new non-classical level of description.

The interpretation of QT which I shall be proposing here is in some sense a half-way house between the (new) orthodox interpretation and the 'realist' interpretation of Einstein *et al.* (although as an interpretation it is totally realist, both in the sense that it interprets the referring terms of QT realistically and endorses metaphysical realism).

I shall retain the (new) orthodox interpretation's notion that within QT reality is described in a radically non-classical way in terms of state-vectors, and more generally in terms of density operators. Moreover I shall follow Born's idea that q-systems are associated with fields, indeed probability fields, which pilot them; an idea which was, I argue, prematurely given up in the face of Einstein's criticisms. I shall take it that these fields are not described in terms of potential functions as they are in classical field theory, or in terms of metrics as they are in general relativity, but in terms of state-vectors or, more generally, density operators. I shall also follow Bohr's notion that q-quantities (i.e. the physical quantities referred to in QT) may be indeterminate in value. Moreover I shall retain in generalized form the orthodox principle that Q is indeterminate in value in S at t if the state-vector for S at t is not an eigenvector of Q. In particular I shall propose the following density operator criterion for q-quantities to have determinate values:

(Bohr) Q in S at t has a determinate value if $\hat{W}(S, t)$, the density
         operator for S at t, is diagonal in the eigenvectors of Q for S.
         (Although note that I do not put forward the converse to
         this. The 'if' here is not also an 'only if'. Indeed to put
         forward the converse leads to a contradiction, as we shall see
         in Chapter 4.)

The retention of these more or less orthodox ideas including Born's
probability field enables the explanation of various quantum
phenomena such as interference effects in just the way that the
orthodox interpretation does. Thus we can (in part at least) explain
the diffraction pattern observed when a stream of electrons is passed
through a double-slit by saying that the position of each electron as it
passes through the slits is indeterminate, and hence does not simply
pass through just the one slit as a particle would.

Where my views diverge from the orthodox interpretation however
is in my rejection of Bohr's notion of complementarity: that (to put it in
Feyerabend's terms) the mode of description appropriate to a
quantum system (whether we describe it as particle, or as wave, say) is
dependent on the experimental conditions, and in particular depends
on what measurements the quantum system is being subjected to.
Instead I take the quantum system always to be a particle (albeit one
with indeterminate values for some of its physical quantities) *and*
always to be guided by a probability field. Thus its wave and particle
aspects are always co-present. This metaphysical difference from the
orthodox interpretation turns out to have important consequences
for explaining the Stern–Gerlach experiment.

I also differ from some orthodox views in putting forward an
interpretation of probabilities in QT as not simply epistemic, and in
particular as independent of our state of knowledge. I also endorse a
non-subjective view of the indeterminacies of QT, making a point of
distinguishing the epistemological term 'uncertainty' from the ontological term 'indeterminacy'. On both of these points I am opposed to
'anti-realist' or 'subjectivist' interpretations of QT, which embrace
reference to subjective entities within the theory.

The criterion (Bohr)' which I mentioned above has several
important consequences. Firstly it enables an interpretation of the
measurement process, which does not assume a reduction (or
'collapse') of the wave-packet. Indeed, on my view, such a collapse

turns out to be inconsistent with the laws of QT. Moreover (Bohr)' has the consequence that on special occasions *all* the q-quantities of a q-system, in particular both its momentum and position, have determinate values, just as the realist says they do. This consequence contradicts the claim made by the orthodox interpretation that incommensurables such as momentum and position never have determinate values simultaneously. By diverging from the orthodox interpretation in this way we can solve two of the paradoxes which are standardly brought to bear against the orthodox interpretation: the Schrödinger cat paradox and the Einstein–Podolski–Rosen (EPR) paradox. The principle (Bohr)' solves these paradoxes by showing that (as the 'realist' claims) in the situations described in those paradoxes *all* of the q-quantities for the relevant systems (the cat and the indirectly measured electron respectively) have determinate values. Thus Schrödinger's cat is rescued from its state of indeterminate morbidity, and the EPR electron has elements of reality corresponding to *both* its momentum and position.

There are however various new problems which the principle (Bohr)' raises. Firstly there is the problem of interpreting the situation in which, although S at t is in an impure state with $\hat{W}(S, t)$ diagonal in the eigenvectors of Q, there is also a more comprehensive system S $+S'$ in a pure state at t with $\hat{W}(S+S',t)$ *not* diagonal in the eigenvectors of Q. This problem will be discussed in detail in Chapters 4 and 5. There are also problems raised by the Kochen and Specker theorem, which, I will show in Chapter 7, forces us to adopt a 'de-Ockhamized' form of representation for q-quantities. Also, in Chapter 9, I shall argue that we should give up the equality between the probabilities of measurements registering particular values and the probabilities of the measured quantities possessing those values.

And there is another problem for (Bohr)' which it shares with hidden variables (and 'realist') interpretations of QT: the problem of non-locality. In solving the EPR paradox, (Bohr)' implies that each q-quantity for both of the electrons in the correlated EPR state has an element of reality, i.e. is determinate in value, a view which hidden variables theories share. But Bell's theorem can be used to show (so it is claimed) that this last result has the consequence that non-local effects are manifested at one of the electrons as a result of an act of measurement performed on the other electron. I shall discuss this objection in Chapter 8. In general terms my solution will be to give up any one-to-one connection between (counterfactually construed) measured values and possessed values for q-quantities to embrace what Redhead calls 'indeterminism'.

If the suggestions I make for resolving the various objections to (Bohr)' are successful then the interpretation of QT which is put forward here will have been justified. It will have been shown to combine the advantages which the orthodox interpretation has in explaining interference effects (a task which the realist interpretation never manages without some awkwardness) with the advantages which the realist interpretation has over the orthodox interpretation in resolving Schrödinger's cat and the EPR paradoxes.[1] In short, the interpretation here will be presented as a minimal modification of the orthodox interpretation in the direction of the realist, a modification which is just enough to resolve the EPR and Schrödinger cat paradoxes but not so much that it generates new (and worse) problems.

[1] Bohr of course claims to resolve the EPR paradox, but even if one accepts his resolution, Schrödinger's cat remains a difficulty unless one adopts Bohr's '*Schnitt*' between the macro and micro domains.

1

Bohr

The wave picture and particle picture; limitations of the particle picture; the double-slit experiment; defending the particle picture; difficulties for the wave picture; the collapse of the wave-packet; Bohr's instrumentalism; the new quantum theory; indeterminacy; degrees of indeterminacy.

INTRODUCTION

Bohr took it as the key fact of QT that neither of the models for describing the behaviour of systems offered by classical physics—neither the 'particle picture' nor the 'wave picture'—fitted q-systems in all aspects of their behaviour.[1] More particularly he took these two pictures to be 'complementary'—when one fitted the other one did not. This failure of fit was seen by Bohr as usually only partial however, as a degree of 'imprecision of fit' rather than a total failure of fit; and it is in terms of this notion that Bohr was then able to give expression to his principle of complementarity, viz. that the degrees of imprecision of fit of the wave and particle pictures are reciprocally related: to the extent that one of the pictures fits well, the other fits badly. In this chapter I shall discuss these notions in the context of a more general discussion of Bohr's philosophy of Quantum Theory.

1. LIMITATIONS OF THE PARTICLE PICTURE

For a q-system which is free (i.e. not subject to any forces)[2] the 'particle picture' consists of particular value assignments to the

---

[1] See Bohr (1934) pp. 59ff. This is a reprint of the 1927 Como Lecture in which Bohr presented his ideas in public for the first time.

[2] The forces referred to here are only those theorized within non-relativistic QT, and do not include gravity for example.

system's spatio-temporal coordinates. In other words the particle picture specifies the q-system's trajectory in space and time. On the other hand the 'wave picture' consists of assignments of particular values to momentum (really the three components of momentum) and energy.[3] Why are the properties of having a particular momentum and energy taken to be wavelike properties? It is because, as we shall see, any q-system which has a particular momentum behaves like a wave, in particular it exhibits a diffraction effect on being passed through a pair of closely spaced slits; and this effect then allows us to associate a wavelength with the q-system. (A different experimental effect, the photoelectric effect, allows us to associate a wave-frequency with any q-system which has a particular energy but we will not be considering that experiment here.) Moreover this same diffraction effect, as exhibited in the 'double-slit experiment', not only allows us to associate a wavelength with a q-system, but also presents us with a case where the particle model for the q-system breaks down. In this way the double-slit experiment presents us with a direct confirmation of Bohr's philosophy of complementarity, i.e. it shows that it is just when a q-system exhibits wavelike properties that the particle picture does not fit.

In the double-slit experiment a homogeneous stream of q-systems, say electrons, is fired at a screen with two closely spaced slits in it so that each electron has zero component of momentum in the plane of the screen, and the same fixed momentum p perpendicular to that plane. We then observe the resulting pattern of electrons landing on a photographic plate parallel to but some distance behind the screen. This pattern will show how passage across the screen modifies the stream of electrons.

The pattern traced out by electrons landing on the plate, which we would expect to observe if electrons were particles, would be basically a two-slit image—an image of slit 1 on that part of the plate opposite slit 1, and a similar image opposite slit 2. (We might expect the images to be a little blurred because the initial momentum of some of the electrons will be randomly modified by their collision with the edges of the slits.)

On the other hand were the stream of electrons a wave phenomenon then the slits would diffract the wave and we would observe a

---

[3] For non-free q-systems in a non-trivial potential field, momentum and energy do not commute and hence cannot belong to the one picture.

characteristic diffraction pattern on the plate: a whole series of parallel images of the slit, not just blurred, but running together separated by penumbra of continuously varying intensity: in short a series of maxima and minima of intensity. From the nature of this pattern we can then determine the wavelength of the wave phenomenon by a well-known formula.[4]

In fact we observe the diffraction pattern—indeed all q-systems with fixed momentum show diffraction effects which depend on their momentum in just this way. We can then deduce via the above-mentioned formula what wavelength λ a stream of q-systems with a given momentum p would have to have, were it literally a wave, in order to produce these diffraction effects. It is this hypothetical wavelength which is taken to be the wavelength associated with each such q-system. And we can then experimentally determine that the wavelength λ is related to the momentum p by the de Broglie relation: $p = h/\lambda$ (where h is Planck's constant.)

I shall now show that, given certain plausible assumptions, the results of a slightly modified version of this experiment refute the particle picture for electrons. In other words it is not just that the results of the double-slit experiment suggest a wave picture but the results are inconsistent with the particle model, given certain other plausible assumptions. Thus we have our promised confirmation of Bohr's complementarity principle. The derivation I shall use initially is somewhat non-standard in that I use relative frequencies instead of probabilities in order to avoid some of the assumptions made in more standard derivations.

In short, I shall now set up a *reductio* argument in which the initial premiss is that electrons, under the experimental conditions of the double-slit experiment, satisfy the particle picture, and in particular they all have a spatial location at all times during the experiment. (It is this premiss which we will end up rejecting, by showing that, together with certain other more plausible assumptions and experimental results, it leads to a contradiction.) We assume also that the spatial locations of particles are continuous in time, and satisfy the minimally Newtonian property that they will not change direction without an external force being impressed upon them. And finally we assume that

---

[4] For the case of a one-dimensional screen the wavelength is given by $d/2 \sin a$, where d is the slit width and $\sin a$ is the sin of the angular separation subtended at the slits between the central maximum of the pattern and an adjacent maximum. For a discussion of this see Heisenberg (1930), ch. 1 for example, or Feynman (1965), ch. 1.

measurements of location are 'faithful', e.g. a particle registers on the photographic plate in the double-slit experiment iff it actually arrives at the point where it is registered, although we do allow that registration on the plate alters the electron's further trajectory.

Now let E_0 be the stream of electrons which are registered on the plate in the double-slit experiment. From our initial assumptions it follows that each member of E_0 has a continuous and determinate trajectory starting on the side of the screen which is remote from the plate. Since each member of E_0 registers on the plate, and since *ex hypothesi* measurements of position are faithful, it follows that each trajectory of the electrons in E_0 terminates at the plate. But a continuous trajectory which starts on one side of the screen and terminates on the other side (at the plate) must intersect with the screen,[5] and it can only do so at S_1 or S_2 since (we assume) the screen is opaque to electrons except at S_1 or S_2. Hence each member of E_0 passes through S_1 or S_2. Moreover since the electrons are minimally Newtonian, and force-free, we see that each electron passes through *only* one slit (i.e. it cannot double back to go through both slits.)

Now consider a new stream of electrons E_1 identically prepared to E_0 except that only the slit S_1 is open when they transit the screen. Experimentally we determine that the pattern on the plate left by those electrons is just the image of one slit opposite S_1.[6] And similarly the pattern on the plate due to yet another stream of electrons E_2, which registers on the plate when only S_2 is open, is the image of one slit opposite S_2. With these additional experimental results we will now prove by *reductio* that the particle picture must break down for at least those times when the electrons in E_0 pass the screen, or at least must do so if we accept the various auxiliary assumptions and experimental results already cited in the argument together with a few more plausible assumptions.

Let 'x' denote the property of an electron being registered in the neighbourhood of the point x on the plate and $rf_1(x)$ be the relative frequency with which electrons in the stream E_1 are registered in the neighbourhood of point x on the plate. Note that I here deal with relative frequencies, not probabilities, in order to avoid questions of

---

[5] By the mean value theorem, applied to each electron's position considered as a continuous function of time.

[6] Actually electrons, like waves, do exhibit single-slit diffraction patterns, but these have a far greater amplitude than the double-slit pattern and hence can be ignored for our purposes.

whether the corresponding probabilities exist. And note also that I implicitly take *finite* neighbourhoods about the point x in order to avoid talking about continuous distributions. Indeed from now on I take 'at x' to mean 'in a small but finite neighbourhood of x'. Let $rf_2(x)$ and $rf_0(x)$ be the correspondingly defined relative frequencies for the streams E_2 and E_0 respectively. Finally let 'S_1' denote the property of passing through S_1 at some time and similarly for 'S_2'.

Since (as shown above) each electron in E_0 which is registered somewhere on the plate either passes through S_1 or passes through S_2 but not both, it follows that the set of members of E_0 which have the property x (of registering at x on the plate) is equal to the set of members of E_0 which have not only the property x but also either the property S_1 or the property S_2 (but not both.) But, by the distributive law of logic, the latter set of electrons is equal to the set of electrons in E_0 which *either* have the properties x and S_1 *or* have the properties x and S_2. But the latter set of electrons is equal to the union of the set of electrons which have properties x and S_1 with the set of electrons which have properties x and S_2.[7] Thus the set of electrons in E_0 with the property x must be equal to the union of two sets, viz. the set of electrons in E_0 with the properties x and S_1 and the set of electrons in E_0 with the properties x and S_2 (where these are clearly non-intersecting sets, since no electron passes through both S_1 and S_2.) We can summarize these last moves as follows:

$$\{e:e \text{ has } x\} = \{e:e \text{ has } x \text{ and } (S_1 \text{ or } S_2)\}$$
$$= \{e:e \text{ has } (x \text{ and } S_1) \text{ or } (x \text{ and } S_2)\}$$
$$= \{e:e \text{ has } x \text{ and } S_1\} \cup \{e:e \text{ has } x \text{ and } S_2\}.$$

Now we apply the principle that the cardinality of the union of two finite sets with null intersection is the sum of the cardinalities of the individual sets. And hence we see that the number of electrons in E_0 with property x is equal to the number of electrons in E_0 with properties x and S_1 plus the number of electrons in E_0 with properties x and S_2. Thus if we let '$rf_0(x \text{ and } S_1)$' be the relative frequency with which electrons in E_0 have x and S_1, i.e. $rf_0(x \text{ and } S_1)$ is the number of electrons in E_0 with x and S_1 divided by the number of electrons in E_0, it follows that

(i) $rf_0(x) = rf_0(x \text{ and } S_1) + rf_0(x \text{ and } S_2)$.

---

[7] Here we use the principle that $\{x:Fx \text{ or } Gx\} = \{x:Fx\} \cup \{x:Gx\}$.

Now following Reichenbach[8] we can assume:

(Ind) Whether or not S_2 is open has no causal relevance to the trajectories of those electrons in E_0 which go through S_1.

(In particular there is no 'action at a distance' which would allow the trajectories of electrons going through S_1 to be affected by whether or not S_2 is open.) And this in turn is taken to imply that whether or not S_2 is open is statistically irrelevant to the trajectory of those electrons which go through S_1, and in particular statistically irrelevant to whether they land at x on the screen. (This follows from a Reichenbachian principle of causation[9] which takes causal irrelevance to imply statistical irrelevance.)

The statistical irrelevance mentioned here can be manifested over the following more sophisticated experimental set-up which we now use to frame the simple experimental set-up so far. An ensemble E of electrons is prepared as for E_0 except that the slit S_2 is opened randomly with a particular probability $P_E(S_2$ open). Our earlier ensembles E_0 and E_1 are then retrospectively taken to have been sub-ensembles of E; in particular E_0 is taken to be that sub-ensemble of E the members of which transit the screen on occasions when S_2 is open, and E_1 to be that sub-ensemble of E the members of which transit the screen on occasions when S_2 is shut. The statistical irrelevance we have just postulated is then manifested as:

$$P_E(x|S_1.(S_2 \text{ open})) = P_E(x|S_1),$$

which is easily shown to imply:

$$P_E(x|S_1.(S_2 \text{ open})) = P_E(x|S_1.(S_2 \text{ shut})).[10]$$

---

[8] See (1944) pp. 24–32.   [9] See Salmon (1975) p. 124 for example.
[10] Let $P(A/B) = P(A)$. Since $P(A) = P(A.B) + P(A.\bar{B})$ it follows that

$$P(A.B)/P(B) = P(A.B) + P(A.\bar{B}).$$

Hence

$$P(A.B)((1/P(B)) - 1) = P(A.\bar{B}),$$

and hence

$$P(A/B) = P(A/\bar{B}).$$

And this whole proof can be repeated after conditionalizing the probabilities on C, to give the theorem that if $P(A/B.C) = P(A/C)$ then $P(A/B.C) = P(A/\bar{B}.C)$. We then let $A = x$, $C = S_1$, and $B = (S_2$ open).

But we have defined E in such a way that

$$P_E(x|S_1.(S_2 \text{ open})) = P_0(x|S_1)$$

$$P_E(x|S_1.(S_2 \text{ shut})) = P_1(x).$$

Hence

(iii)a $P_0(x|S_1) = P_1(x)$, i.e. $P_0(x \text{ and } S_1) = P_1(x)P_0(S_1)$.

A similar argument can be given for S_2 so that we can show:

(iii)b $P_0(x \text{ and } S_2) = P_2(x)P_0(S_2)$.

Note that the probabilities equated in (iii)a and (iii)b are the *empirical* probability $P_1(x)$ (which we can estimate via $\text{rf}_1(x)$), and what we might call 'a *metaphysical* probability' $P_0(x|S_1)$ which we cannot directly observe in the given experimental set-up. (That is to observe $P_0(x|S_1)$ via its corresponding relative frequency we would need to check which slits particular electrons pass through, but that would change the experimental set-up in such a way as to destroy the diffraction effects. For a good discussion of this see Feynman (1965) ch. 1.) It is of course only because we are assuming the particle model here that the latter probability (or indeed the corresponding relative frequencies) can be assumed to exist.

Now we equate the probabilities in (iii)a and (iii)b with the corresponding relative frequencies, to give:

(iii)c $\text{rf}_0(x \text{ and } S_1) = \text{rf}_1(x)\text{rf}_0(S_1)$

(iii)d $\text{rf}_0(x \text{ and } S_2) = \text{rf}_2(x)\text{rf}_0(S_2)$.

The equation of probabilities with relative frequencies may be justified either directly by adopting a frequency interpretation of probabilities, or by adopting a single-case propensity interpretation of probabilities and appealing to the law of large numbers. (In the second case the identities are of course only approximate, but that is good enough for our purposes here. This issue of the interpretation of probabilities will be discussed in Chapter 2.)

Substituting (iii)c and (iii)d into (i) then gives:

(iv) $\text{rf}_0(x) = \text{rf}_1(x)\,\text{rf}_0(S_1) + \text{rf}_2(x)\text{rf}_0(S_2)$.

But it is precisely (iv) which is experimentally falsified by the diffraction pattern produced by the electrons in E_0 impinging on the plate, i.e. (iv) implies that there should be a simple two-slit image on the plate, whatever $rf_0(S_1)$ and $rf_0(S_2)$ are.[11] But experimentally there is seen to be a complex diffraction pattern. So we have our contradiction, and the particle picture for q-systems, given all the other assumptions made in the above derivation, is seen to be experimentally refuted.

My use of relative frequencies in the above derivation instead of probabilities may seem somewhat strange. One could recast the whole derivation in terms of probabilities as follows (and this is the more usual procedure). One starts with the premiss:

(i)′ $P_0(S_1 \text{ and } S_2) = 1.$

(This is taken to be a consequence of the particle model that each particle which registers on the plate passes through either S_1 or S_2.) Hence, by probability theory,[12] it follows that

$$P_0(x) = P_0(x \text{ and } (S_1 \text{ or } S_2)),$$

where P here indicates a probability, in distinction to rf which indicates the corresponding relative frequency. Hence

$$P_0(x) = P_0((x \text{ and } S_1) \text{ or } (x \text{ and } S_2))$$

(by the distributive law of logic). And this

$$= P_0(x \text{ and } S_1) + P_0(x \text{ and } S_2) - P_0((x \text{ and } S_1) \text{ and } (x \text{ and } S_2))$$

(by the additivity law for probabilities).

But we can take $P_0((x \text{ and } S_1) \text{ and } (x \text{ and } S_2))$ as 0, since S_1 and S_2 are mutually exclusive.[13] Hence this

$$= P_0(x \text{ and } S_1) + P_0(x \text{ and } S_2).$$

But by the independence assumption ((iii) above) this

$$= P_1(x)P_0(S_1) + P_2(x)P_0(S_2).$$

And it is only at this stage of the derivation that we substitute probabilities with their corresponding relative frequencies, and hence derive (iv), thus refuting the particle picture as above.

[11] For non-zero $rf_0(S_1)$, $rf_0(S_2)$ at least. And if one of them is zero, so that the other one is 1, then (iv) implies that there should be a single-slit image—and that too is experimentally falsified.

[12] i.e. it is a theorem of probability theory that if $P(A) = 1$ then $P(B \text{ and } A) = P(B)$.

[13] i.e. it is a consequence of the particle model that the one electron cannot go through *both* S_1 and S_2, provided that we do not allow doubling back of the electrons.

This second derivation has important differences from the first derivation. In particular the second derivation uses more logic and probability theory than the mixture of set theory and arithmetic used in the first derivation. It is also important to note that the first derivation does not need the particular probability assumptions:

$$P_0(S_1 \text{ or } S_2) = 1, \ P_0(S_1 \text{ and } S_2) = 0,$$

which, although they can be made plausible enough,[14] act as additional points at which criticism can be directed by those who wish to defend the particle picture.

2. DEFENDING THE PARTICLE PICTURE

The first (or indeed second) derivation in the previous section is vulnerable at several points, points which a defender of the particle picture may seize upon. For example although Popper accepts that both $P_0(x|S_1)$ and $P_1(x)$ exist, he questions their equality as asserted in (iii),[15] i.e. he questions the independence assumption (Ind). This possibility is dismissed by Reichenbach because it allegedly introduces a causal anomaly.[16] However Landé has suggested a mechanism for the failure of (iii) which clearly involves no such anomalies.[17] And the mere *possibility* of a mechanism such as Landé's invalidates Reichenbach's claim.[18]

There are however more serious problems for Popper's suggestion (of (Ind) failing) than those posed by Reichenbach. (Ind) gains some indirect support from an earlier range of experiments performed by

---

[14] Although it should be emphasized that the step from the premiss 'No electrons go through both slits' to '$P_0(S_1 \text{ and } S_2) = 1$' is by no means trivial, even if one can strengthen the premiss to some suitable construal of 'No electrons *can* go through both slits.' At issue here is the difficult question of the connection between modal and statistical discourses, which we will return to at various points later.

[15] See Popper (1982), sect. 18.   [16] See Reichenbach (1944).

[17] Landé (1965) develops a theory according to which the geometrical shape of the screen—and in particular whether or not the slit S_2 is open—determines how much momentum the screen can transfer to particles colliding with it, and in particular affects particles passing through S_1.

[18] Landé's mechanism is however not all that plausible. It has the drawback of not being generalizable to explaining other cases of interference effects—such as the Stern–Gerlach experiment. Also its critical dependence on factors such as geometrical shape make it a poor candidate for a relativistic generalization. Its plausibility is not at issue here however.

Mandel and Pfleegor (1968) for example, and more lately by Grangier, Roger, and Aspect (1986).[19] These experiments showed that interference effects occur between two photon beams even when the photon flux is so low that the photons in one beam may be considered (on the particle model for photons) to be clear of the apparatus whenever there are photons from the other beam present. As such no causal action of the one beam on the other could explain the interference effects if we adopt the particle model for photons. Moreover it is hard to see how Landé's mechanism could explain the interference effects since there is no macroscopic apparatus in the region passed through by both beams. This experiment indicates that dropping (Ind) is of no general help in reconciling the particle model with the existence of interference effects even if it does provide a possible explanation in the special case of the double-slit experiment. And this in turn undermines the rationale for dropping (Ind) in the case of the double-slit experiment as well. That is if we need a hypothesis, other than the falsity of (Ind), for reconciling the particle model with the existence of interference effects in contexts other than the double-slit experiment, then that same hypothesis, whatever it is, will probably do for the double-slit experiment.

A second point at which a defender of the particle model could block the first derivation of (iv) is at the apparently uncontroversial argument for (i), e.g. he could question the principle that $\{x:Fx \text{ or } Gx\} = \{x:Fx\} \cup \{x:Gx\}$. Or, following Putnam's suggestion in (1969), he could drop the distributive law of logic. Or of course the experimental results themselves could be questioned. (It is interesting to note that notwithstanding certain strong versions of the Quine–Duhem thesis, which assert the malleability of the empirical data, this last option is never considered.)

In short we see that there are several ways of retaining the particle model in the face of the above experimental refutation of (iv). But all these ways put some strain on the framework of assumptions within which a particle model is usually advanced. There is therefore some justification for saying with Bohr that the double-slit experiment shows the failure of the particle model, despite the existence of these alternatives.

Moreover, in favour of this last conclusion, I point out that there is yet another way of arguing for the conclusion that the results of the

[19] For an early review see Pipkin (1978).

double-slit experiment contradict the particle model, a way which uses *modal* terminology instead of the statistical terminology used above. I mention this third way here not only because it circumvents some of the attempts to salvage the particle model just mooted but also because it introduces the distinction between statistical and modal approaches to causation, a distinction which will be of importance in the final chapters (where we discuss the inference from statistical correlations to certain modally construed causal anomalies).

Reconsider the ensemble E_0 of electrons. Let x be one of the central null points of the diffraction pattern on the plate, i.e. *no* electrons from E_0 register at x although x is central to the pattern (near one of the central peaks).[20] And let $E_0(1)$ be the sub-ensemble of E_0 for which the members all pass through S_1, and similarly define $E_0(2)$. Clearly $E_0 = E_0(1) \cup E_0(2)$; and hence, since *ex hypothesi* E_0 has a large number of electrons in it, so must at least one of $E_0(1)$ or $E_0(2)$. Let it be $E_0(1)$. We now make the 'independence assumption' (Ind)' that for any member e of $E_0(1)$ were slit S_2 shut as e crosses the screen (instead of being open as we suppose it to be in fact) then e would still register at the same place on the plate; or more formally:

(Ind)′ For any location x' on the plate and any electron e in $E_0(1)$, if e registers on the plate at x' then were S_2 shut as each of the electrons in $E_0(1)$ crosses the screen e would still register at x' on the plate.

Here we have a modal construal of the independence assumption (that S_2 being open or shut does not affect electrons going through S_1), as opposed to the statistical construal given previously. Note that for convenience I shall here analyse the counterfactual 'Were p then it would be the case that q' as simply 'In *the* nearest possible situation to the actual for which p takes place so does q'.[21]

Now consider the nearest possible situation to the actual at which S_2 is closed while the various members of $E_0(1)$ cross the screen. And let $E_0'(1)$ be the counterpart of $E_0(1)$ for that possible situation. Since

---

[20] A complication arises here. It will be remembered that we stipulated 'at x' to mean 'in a small but finite neighbourhood of x'. We now see that these neighbourhoods have to be small enough that effectively *no* electrons land in the neighbourhood of one of the central 'null' points of the diffraction pattern.

[21] I generalize this in later chapters. See Lewis (1973) for a discussion of possible world analyses of counterfactuals.

QT still holds for that possible situation,[22] it follows that the pattern on the plate made by the members of $E_0'(1)$ is the single-slit pattern, given by the above probability function P_1. But $P_1(x) \gg 0$ since x is a central point (and despite $P_0(x) = 0$). Hence, assuming that probabilities at least approximate the corresponding relative frequencies over large ensembles, it follows that there will be at least *some* electrons in $E_0'(1)$ which register at the point x (near the central peak). Let e be one such electron.

From (Ind)' we can then trivially prove that in the actual world some electrons in $E_0(1)$ must also register at x, as follows. First we remark that e has a counterpart which belongs to $E_0(1)$ in the actual world since $E_0'(1)$ was defined to consist of counterparts of the very same electrons which are in $E_0(1)$. Now let x' be the point on the screen in the actual world where the counterpart of e registers. Hence, by (Ind)', were S_2 shut as the electrons in $E_0(1)$ cross the screen then e would register at x'. But this means that e registers at x' in the nearest possible world to the actual at which S_2 is shut as the electrons in $E_0(1)$ cross the screen. But *ex hypothesi* e registers at x in the nearest possible world to the actual at which S_2 is shut as the electrons in $E_0(1)$ cross the screen. Hence $x = x'$. And hence as required we have shown that in the actual world some electrons in $E_0(1)$ (in particular the counterpart of e) register at x.

Moreover we see that we have derived a contradiction. That is we have just shown that there are some electrons in $E_0(1)$ which register at x in the actual world, but earlier on we denied precisely this, since x was defined as a null point of the two-slit diffraction pattern. Thus the experimental fact that there are such null points (which are not null points of the single-slit pattern) provides a point of contradiction with the particle picture together with the various other assumptions made in the derivation, viz. the independence assumption (Ind)', and the various logical and set theoretic principles.

This contradiction with experiment means that we must face one of the following options: to give up the particle picture, or the independence condition (Ind)' (now a modal condition), or one of the logical or set theoretic principles used in the derivation (always assuming of course that we accept the experimental results). But giving up the independence condition poses exactly the same

[22] The question of the domain of possible situations in which the laws of QT hold will be dealt with in detail in Chapter 8.

problems here as it did in section 1. So again giving up the particle picture is the preferred option.

As a basis for criticizing the particle model this last derivation does have considerable advantages over the earlier derivations. (My reason for giving the earlier derivations nevertheless is that some may find the use of counterfactuals in this last derivation objectionable.) The advantages are that various of the auxiliary assumptions—such as the existence of the joint probabilities $P_0(x$ and $S_1)$ and $P_0(x$ and $S_2)$, and even (it would seem) the law of distribution—used in the earlier derivations are no longer needed in this last derivation; and hence various of the ways—such as Fine's (which rejects the existence of the relevant joint probabilities),[23] and Putnam's (rejecting the distributive law)—of keeping the particle model are shown to be inadequate by this last derivation.

(In defence of Putnam however, it must be conceded that the derivation just given does make use of logical apparatus which it can be argued implicitly includes the distributive law of logic, i.e. the last derivation essentially started from the disjunctive premiss that *either* $E_0(1)$ *or* $E_0(2)$ has a large number of members, derived a contradiction for each of these alternatives with the help of various side premisses, and then concluded that the initial disjunction is in contradiction with the relevant side premisses. But this pattern of argument is tantamount to the acceptance of the principle of or-elimination.[24] And if we assume or-elimination, commutation of disjunction and conjunction, and the two usual inference rules:

p therefore (p or q),
(p and q) therefore p,

then the law of distribution can be easily shown to follow. So Putnam's thesis—that we can save the particle model in the face of the double-slit experiment by rejecting the law of distribution—is preserved, albeit in a somewhat more complex and perhaps less plausible form.)

It is important to note that the account of the double-slit experiment given in this section deals only with a *single* system, by contrast with the account of the previous section which dealt with an ensemble of systems. It is the use of a modal rather than statistical

---

[23] See A. Fine (1971).   [24] See Lemmon (1965) p. 22.

interpretation of the independence condition which makes this change possible. One consequence of this change is that the wavelike properties of electrons—or, more narrowly, their non-particle-like properties—are seen to be not just emergent properties of groups of electrons. In other words it is not just that groups of electrons behave like waves, it is *also* the case that individual electrons do not behave like particles. This is clear from the fact that when one slit is closed some electrons go where they could not go were both slits open. In other words the non-particle-like behaviour is manifested not just statistically, over a whole ensemble of electrons, but in the behaviour of an individual electron, viz. the electron e which happens to register at x (a null point of the two-slits-open interference pattern). So it seems that we cannot dismiss the interference effects as a mere co-operative phenomenon and thereby retain the particle model for individual electrons.

Finally in this section I note that it is tempting to modify the last derivation to provide a swift and direct experimental refutation of the law of distribution as follows. From the premisses of the first part of the derivation it follows that for *any* electron e in E_0 '$(x(e)$ and $(S_1(e)$ or $S_2(e))$)' is false, where '$x(e)$' abbreviates 'e lands at x' and '$S_i(e)$' abbreviates 'e passes through S_i'. But from the premisses of the second part of the derivation it follows that for some e in E_0, '$(x(e).S_1(e))$' is true, and hence '$(x(e).S_1(e))$ or $(x(e).S_2(e))$)' is true too. Thus the law of distribution fails. QED. But this quick argument derives the failure of distribution from premisses which (we have shown) are already in contradiction with each other, a contradiction which must surely be removed whether or not one also rejects distribution. But to remove this contradiction means rejecting one of the premisses in the derivation, and in rejecting one of them (whatever it is) the proposed modification of the derivation into a direct argument for the failure of distribution must also be rejected.

3. DIFFICULTIES FOR THE WAVE PICTURE

In the previous sections we showed the difficulties involved in sustaining a particle picture for a q-system (in particular in assigning it a definite spatio-temporal trajectory), difficulties manifested in the double-slit experiment. A quite different experimental situation can be constructed which shows that for a different q-system (one in a superposition of definite momentum states) the wave picture (i.e. the picture assigning particles a definite momentum, and hence wavelength—see section 1) runs into difficulties.

Suppose the following three experiments are performed. First fire the members of a set of electrons E_1 along a certain line l, each electron having a precise momentum p_1. The electrons are free so that momentum is conserved, but the whole experimental arrangement is taken to be enclosed by walls at a distance L from each other along the line l. Let R be a spatial interval along the line l, and $P_1(R, t)$ be the probability of finding electrons from E_1 in R at time t. It is then easy to prove from QT that $P_1(R, t)$ has exactly the value one would expect were the electrons classical particles with totally uncertain positions, i.e. it has the value $L(R)/L$ where $L(R)$ is the length of R.[25]

Now repeat the experiment, but giving all the electrons in some new set E_2 the momentum p_2 in the same direction along l (where $p_2 \neq p_1$). Clearly $P_2(R, t) = L(R)/L$.

Finally let each of the electrons in some set E_0 be in a state at time t which is a 'linear superposition' of the corresponding states for E_1 and E_2 (with equal superposition coefficients for each state). I shall discuss this notion of a linear superposition of states in detail later. For now I need only note that we can straightforwardly prove (within QT) that

(v) $P_0(R, t) = (L(R)/L) - (\sin(k_1 - k_2)a/(k_1 - k_2))$,

where $k_i = p_i/\hbar$, $\hbar = h/2\pi$, and $a = L(R) - L$.[26]

But this result can be argued to contradict the assumption that the electrons always have a particular momentum (i.e. that the wave picture is universally applicable) as follows. If the electrons always have a particular momentum, and the measurements are faithful, then there is probability $1/2$ that the electrons in E_0 have momentum p_1 and probability $1/2$ that they have momentum p_2. (This is because the probability, derivable from QT, that they are *measured* to have momentum p_1 is $1/2$, and similarly for p_2.)[27] Hence we prove that

$$P_0(R, t) = P_1(R, t)/2 + P_2(R, t)/2$$
$$= L(R)/L,$$

which contradicts (v). Therefore similar difficulties to those which in

---

[25] See Schiff (1955), sect. 11.   [26] See Schiff (1955), sect. 11.
[27] We again assume faithfulness of measurement here.

the case of the double-slit experiment confronted the assumption that *q*-systems always have particular positions here confront the assumption that *q*-systems always have particular momenta.

It is important to note a confusion which sometimes appears in the literature. For Bohr, and more generally for the orthodox interpretation of QT, the wave picture assigns to a *q*-system a definite momentum, and hence (as indicated in section 1), a particular wavelength. This does not however imply that *q*-systems are literally waves, coincident with what we shall later identify as the *q*-systems' state-functions. Indeed the objections to taking *q*-systems literally as waves (as opposed to merely satisfying a wave picture) go well beyond pointing out the wave picture's inability to describe *q*-systems in situations such as we have just outlined above in this section. One such objection, based on the 'collapse (or reduction) of the wave-packet' phenomenon, will be of concern later. For convenience I shall introduce it here.

Consider a single electron fired towards a photographic plate some distance away. Suppose the electron is literally a wave coincident with the electron's state-function. Then it will spread out by the time it reaches the screen. Indeed it can be made to spread as far as we like by putting the plate far enough away. But the electron registers at the plate essentially as a single point. Therefore in the small interval of time it takes to register on the plate it contracts from as far away as we like to essentially a point. This phenomenon is called 'the collapse of the wave-packet'. It requires that the leading edge of the wave travels with a speed which can be made as large as we like simply by arranging for the plate to be far enough away from the point of origin of the electron. But this (so the argument goes) contradicts the law of special relativity that no speed can exceed that of light, at least not for systems with non-zero mass. So the electron cannot literally be a wave, at least not in the sense of having its mass literally spread out to coincide with the boundaries of the wave.

Whatever the merits of this last argument may be (I shall criticize it in Chapter 3) it is at least clear that it does not prevent us from 'associating' a wavelength with the electron in the way discussed earlier. It is only if the electron is literally a wave, with a wave dynamics, that this last argument applies. In short the collapse of the wave-packet, unlike the argument given at the beginning of this section, provides no reason for not applying the wave picture to electrons.

4. BOHR'S INSTRUMENTALISM

We have so far discussed the sort of evidence which can be marshalled in support of Bohr's claim that neither the classical wave picture nor the classical particle picture is universally applicable to q-systems: each applies only in certain contexts to which the other picture ('complementary' to it) does not apply. What may seem puzzling is why, given this claim, Bohr persisted in taking classical pictures as the basic stock of descriptions for q-systems, and this may seem particularly puzzling given the further fact that there are some conditions under which no classical pictures will fit a q-system.[28]

Bohr's reason for persisting with classical pictures seems to have been something like this: we must describe q-systems (or any systems, for that matter) in the terms we use for describing the macroscopic world, because our experience of the macroscopic world is what is used to teach us our descriptive language. And these terms are just the terms of classical physics, which we know to be valid at the macro level. Thus he writes:

It is decisive to recognise that, *however far the phenomena transcend the scope of classical physical exploration, the account of all evidence must be expressed in classical terms.* The argument is simply that by the word 'experiment' we refer to a situation where we can tell others what we have done and what we have learned and that, therefore, the account of the arrangement and the results of the observations must be expressed in unambiguous language with suitable application of the terminology of classical physics.[29]

So how did Bohr accommodate this requirement to use classical pictures in QT, despite their failure to fit q-systems? He did so by taking it to be the role of QT to codify the degrees of imprecision (to be defined in section 6) with which classical descriptions fitted q-systems. In particular there was for him no further question of developing a new non-classical description of q-systems which did fit them precisely.

This view of the role of QT was a natural extension of nineteenth-century 'positivist' views which saw it as the task of physics to

---

[28] e.g. when a q-system is in the superposition of just two momentum eigenstates (as described in the previous section) it does not have a particular momentum; but it also then exhibits diffraction effects (when subjected to a double-slit experiment) so that it cannot be said to have a particular location either.

[29] (1949) p. 209.

construct theoretical models for systems and investigate empirically how well they applied to particular phenomena.[30] Bohr's extension of this view was simply to make the question of the applicability of models a theoretical not empirical question, to be investigated by the use of thought experiments, as exemplified in his debates with Einstein.[31]

One could criticize the empiricist account of concept formation in science which seems to lie behind Bohr's views here. But it will suffice to point out that the subsequent development of QT has refuted Bohr's views. It has refuted them by developing new non-classical concepts which do fit q-systems precisely at all times, although Bohr's talk of 'wave pictures' and 'particle pictures' is retained at a secondary level, as is Bohr's notion that these pictures are 'complementary'. Thus the 'new QT'—developed by Born, Heisenberg, Schrödinger, *et al.*—assigns state-vectors (or equivalently state-functions) to q-systems at *all* times. Moreover this is a non-classical means of describing q-systems in the sense that it does not conform with the classical method of description in terms of a physical quantity being assigned some numerical value.

Bohr's reaction to this new formalism for describing quantum systems was to dismiss it as merely 'instrumental', as not descriptive of anything real: 'The mathematical formalism of quantum mechanics . . . merely offers rules of calculation'.[32]

But, as we shall argue in Chapter 2 *contra* Bohr, to identify the state-vector S at t does describe a level of reality, moreover a level of reality which is 'deeper' than Bohr's classical descriptions ('deeper' because, as we shall see, from the state-vector of S at t we can derive which of Bohr's pictures is appropriate for S at t). And this conclusion is borne out historically by the fact that even the first generation of Bohr's followers showed a tendency to take the emerging formalisms of the new quantum theory (either that of matrix mechanics or wave mechanics) as describing some new level of reality, even if that reality was not being described in a totally 'realistic' fashion by the new

---

[30] See Krips (1985).   [31] See Bohr (1949).

[32] Bohr (1934) p. 60. Note however that although Bohr was instrumentalist about the vector formalism of QT, he was not an out-and-out instrumentalist about QT. In particular he was quite realist about atoms, electrons, etc. Thus his anti-realism was of a variety specifically directed against the mathematical structures postulated within the new QT. In other words he was sceptical only about whether these particular mathematical structures represented any real structures, not about the atomic entities which were supposed to be the bearers of the structures.

formalism. For example Heisenberg writes:

> I had the feeling that, through the surface of atomic phenomena, I was looking at a strangely beautiful interior, and felt almost giddy at the thought that I now had to probe this wealth of mathematical structures nature had so generously spread out before me . . . If nature leads us to mathematical forms of great simplicity and beauty— . . . coherent systems of hypotheses, axioms, etc.— . . . we cannot help thinking they are 'true', that they reveal genuine features of nature (1971, pp. 59 and 68).

The realism expressed in this last quotation is not unequivocal, but neither is it the confident instrumentalism of Bohr. And in a similar equivocal vein Born writes:

> These are structures of pure thinking. The transition to reality is made by theoretical physics, which correlates symbols to observed phenomena . . . these very structures are regarded by the physicist as the objective reality. . . . This procedure leads to structures which are communicable, controllable, hence objective. It is justifiable to call these by the old term 'thing in itself'. They are pure form, void of all sensual qualities. . . . But that they are perfectly empty does not fit the facts. Remember what practical use can be made of them (1949, pp. 227 and 232).

It is important to realize however that this non-classical description of q-systems in the new QT does not *replace* the classical style of description of the old QT. Bohr's classical descriptions have merely been supplemented with a stock of non-classical descriptions referring to a deeper reality. Thus talk of momentum, position, energy etc. and hence of wave pictures and particle pictures survives in modern QT, which means that Bohr's notion that these pictures failed to fit under certain conditions has survived too. And it is to a philosophical discussion of this notion of failure of fit—and more generally of indeterminacy—that I now turn.

5. BOHR AND HEISENBERG ON INDETERMINACY

In the context of the double-slit experiment Bohr ended up by suggesting not that the electrons have *no* spatial location as they cross the screen, but rather that their spatial location is *indeterminate*, which, as Bohr puts it, means that there is a 'degree of imprecision' or 'lack of sharpness' in the very definition of the concept of spatial location for the electrons, an imprecision inherent in the nature of the experimental set up. Thus, in his 1927 Como lecture, Bohr says that for quantum phenomena there is 'a definite latitude in our (classical) account', and 'a limitation of the classical concepts', and that 'an unambiguous definition of the state of the system is naturally no longer possible'.[33]

The same notion of indeterminacy (*unbestimmtheit*) can be found in the early Heisenberg, his favoured term being '*ungenauigkeit*' ('imprecision' or 'inexactness') rather than 'indeterminacy'.[34] This difference of terminology reflected a serious difference between Heisenberg and Bohr. As Jammer points out in section 3.2 of (1974) they differed about the reason behind the indeterminacy. Heisenberg, it seems, located it in an absence of measurement; thus it is because the position of the electron as it passes through the slits is only measured to within the intervals occupied by the slits that its position cannot be defined with a greater degree of precision. Bohr on the other hand never accepted this explanation of indeterminacy in terms of empirical meaninglessness (i.e. a lack of appropriate measurements). Thus we have the two contrasting views—Heisenberg's:

If one wants to clarify what is meant by 'position of an object' . . . he has to describe an experiment by which the position . . . can be measured; otherwise this term has no meaning.[35]

and Bohr's:

The reciprocal uncertainty which always affects the values of those quantities is essentially an outcome of the limited accuracy with which changes in energy and momentum can be defined [not measured].[36]

Nevertheless both Heisenberg (initially) and Bohr conceived of indeterminacy as a lack of precision in applying classical concepts. They disagreed (according to Jammer) because Heisenberg explained this indeterminacy (on verificationist principles) by a lack of measurement, whereas Bohr took the indeterminacy to be the cause of lack of measurability, i.e. Bohr thought that no precise value can be obtained on measurement only because there is no precise value to be measured, this fact in turn resulting from the fundamental wave–particle duality.

[33] In 1925, as Jammer points out p. 91 ff (1974), Bohr had not yet formulated this view, claiming instead 'quantum mechanics does not deal with a space-time description'.

[34] Jammer (1974) p. 61.   [35] See Heisenberg (1930) p. 15.

[36] Jammer (1974) p. 69, my parentheses.

However it must be added here that Heisenberg (1930) p. 1 makes it clear that verificationism has a *heuristic* role only for him in just the way that operationalism does for Einstein. Thus he writes: 'It seems necessary to demand that no concept enter a theory which has not been experimentally verified . . . Unfortunately it is quite impossible to fulfil this requirement, since the commonest ideas and words would be excluded.' And on p. 14 he writes in a most non-verificationist mode, which foreshadows his later writings: 'The indeterminateness is to be considered as an essential characteristic of the electron, and not as evidence of the inapplicability of the wave picture.'

Indeed here we have a specific gesture towards Heisenberg's later view of q-systems as a site for *potentia* (see his (1958) p. 160). So even in the early Heisenberg alleged verificationist arguments are not embraced unequivocally. (On this point see Jammer (1974, p. 69.)

In this book I propose to take a position which corresponds to neither the early Heisenberg's alleged verificationist tendencies nor Bohr's alleged instrumentalism, but is in some respects half-way between them. I agree with Bohr that 'indeterminacy' is not merely due to lack of measurement. Indeed I shall later discuss how the new QT provides a criterion for the indeterminacy of certain properties of q-systems, a criterion expressed in terms of the theoretical categories introduced by the new QT, and which is quite independent of what is measured.[37] But I disagree with Bohr's instrumentalist view of the new quantum formalism. This compromise position can be seen as an explication of the later Heisenberg's views (although I differ from the later Heisenberg in my reading of the density operator formalism, and account of the measurement process).

6. INDETERMINACY

What is this concept of indeterminacy, *qua* imprecision of fit, which Bohr and Heisenberg both took for granted? In this section I shall argue that this concept, as it occurs in QT, can be seen as at one with the traditional concept of indeterminacy to be found in the philosophical literature.

---

[37] In particular the physical quantity Q for the q-system S at time t is taken to be indeterminate in value if the state-vector of S at t is not an eigenstate of Q. Note that this will be so even though a measurement of Q in S at t is taking place.

Certain properties (or property terms) like 'many', 'tall', 'about six feet tall', 'yellow' are traditionally said to be 'indeterminate' (or 'vague' or 'fuzzy' or even just 'open textured'); and this claim is traditionally linked to there being no correct way of either applying or not applying those properties in certain borderline cases, where this lack cannot even be rectified in principle by attention to the wider context in which the terms appear (as it can with properties which are merely ambiguous).[38] For example we cannot say whether or not there are many coins in my pocket given that there are just ten of them. Certainly one—or even two—coins would not count as many (although for some purposes it could be counted as too many), and a thousand coins would be counted as many (although for other purposes it might be too few). Nevertheless the criteria we have for the truth conditions of 'there are many coins in my pocket' leave open its truth-value in the case for which there are just ten coins in my pocket, and they do so whatever further information we may adduce, so that the openness of the truth-value cannot be seen as just a matter of our ignorance.

Here we must introduce a difficulty however. Consider the property of having a particular area. This seems to be a perfectly well-defined, and in that sense 'determinate', property. In particular there seems no room for sharpening up the definition—for 'precisification'—of the concept of having a particular area. Nevertheless there are cases of objects with an indeterminate area (e.g. puddles, battle-zones, cities, etc.) which are on the borderline between having one area or another. In short there is a different notion of indeterminacy at work here for which a property is not indeterminate and yet has borderline cases.

The traditional sources of indeterminacy of properties may be several. For example it may be lack of detailed enough definitions, i.e. conceptual underdetermination, as in the case of 'many', so that we have a truth-value gap whatever our informational state. But it may also arise because definitions are too detailed (conceptual over-determination), so that the criteria *both* for having the property and having some contradictory property are satisfied, given the information we have. For example there are both physiological and

---

[38] The issue of whether to take indeterminacy as a feature of descriptions, or rather of the things/properties/states of affairs described is one I shall not discuss here. See Dummett's comments (1975) p. 314, and for an opposing view see his (1979) p. 9. Also see K. Fine (1975).

genetic criteria for sex, and in certain possible (and actual) cases where these conflict sex is said to be indeterminate.

But indeterminacy may also afflict *objects* (or object descriptions), i.e. it may be objects rather than their properties which are indeterminate. Of course the indeterminacy of an object may simply be due to the indeterminacy of those property terms which are used to single it out. Thus 'the set of yellow objects on the table' may be indeterminate in its reference because one of the objects on the table— say the pen—is borderline yellow-orange. But objects may also be indeterminate in their own right it seems. Thus the city of Minneapolis (Minnesota) is indeterminate, not just because it is indeterminate in spatial extent (after all *any* city is that), but because it is not clear whether we are to distinguish the city of Minneapolis from the neighbouring city of St Paul. Do we have the one city St Paul-Minneapolis (as we now have the one city of Budapest instead of the two separate cities of Buda and Pest) or do we have two distinct cities of St Paul and Minneapolis (as Buda and Pest were once separate cities)?

Note that this indeterminacy of objects may have to be considered as an indeterminacy which is 'under a description'. Thus Melbourne (unlike St Paul) is determinate *as* a city in the sense that it is clearly distinguishable from any other city, but it is indeterminate as a collection of buildings, i.e. it is not clear which buildings belong to Melbourne and which are merely on the outskirts. This suggests that the indeterminacy here is linguistic in nature, i.e. is properly assigned to objects under descriptions rather than objects as such. Fortunately I do not need to confront this issue—the linguistic nature of indeterminacy—but content myself with making the more modest point that mere indeterminacy in spatial extent had better not be taken as sufficient for an entity to be indeterminate or indeterminate under all descriptions). To deny this would mean that all cities were indeterminate, and that conclusion in turn would undercut a distinction we seem to want to make between the indeterminacy of Minneapolis, because of its special relation with St Paul, and the indeterminacy which any city has because of its indeterminate area.

There are yet other examples of indeterminacy, viz. propositions for which neither the subject nor the predicate terms refer to indeterminate objects/properties, but it is nevertheless indeterminate whether that particular subject satisfies that particular predicate. For example it is at least to some extent indeterminate what area the city of Melbourne has: it is indeterminate whether the area is 70 square miles or merely 69 square miles. But arguably this is not because the city of Melbourne is indeterminate as a city. On the contrary Melbourne, *as* a city, is perfectly determinate—perfectly distinct from Sydney, Brussels, London, and all other cities—although it does of course have a somewhat indeterminate location. (This determinacy of Melbourne *as* a city is of course a contingent fact about Melbourne which could be otherwise. For example if the outer suburbs of Melbourne and the city of Warragul, premier city of Gippsland, grow into each other sufficiently then it may well become indeterminate what the city of Melbourne is, not just where it is, e.g. it will be indeterminate whether Warragul-Melbourne is one or two cities, whether Warragul takes on the properties of Melbourne, and vice versa, etc.) Nor, it seems, does the indeterminacy here arise because predicates of the form 'has area *A*', for various numbers *A*, are indeterminate. As we have indicated already, the property of having a particular area is perfectly determinate in the sense of 'well defined'; it just happens that Melbourne's location is not determinate enough to allow the definition of a particular area for it.

In short it seems that the indeterminacy of Melbourne's area arises because, so to speak, the family of predicates 'has area *A*' constitutes a conceptual grid into which the (nevertheless well-defined) entity 'Melbourne' does not fit. We may further illustrate this notion of indeterminacy through failure to fit a conceptual grid by considering the eponymous *platypus* (flat-foot). Historically the platypus did not fit those eighteenth-century zoological classificatory schemes which opposed 'oviparous' to 'mammalian', because the platypus satisfies some, indeed most, of the criteria for mammal-hood, including being warm-blooded and suckling its young (although not the criterion of giving birth to its young) but also satisfies the central (defining, one may say) characteristic of being oviparous (viz. laying eggs). For this reason its historical status—is it mammalian or not—was in our terms 'indeterminate'. Moreover note that the indeterminacy here was *not* due to a failure of determinacy of the individual properties for being mammalian, *nor* a lack of determinacy of what it was to be a platypus. Rather it is because the platypus did not fit the relevant classificatory scheme. (It is interesting to note that historically the zoological classificatory scheme was shifted in order to excise this indeterminacy once attempts to discredit the existence of the boundary-crossing platypus had failed—see Gould's comments

(1983) in his *Mammals of Australia*. By contrast, in the case of QT, such indeterminacies were incorporated into the theory rather than excised.)

In the case of Melbourne's indeterminate area one can locate a similar 'failure of fit' between the entity Melbourne and the conceptual grid consisting of various assignments of areas. Ultimately this indeterminacy arises because certain locations on the outskirts of Melbourne have some of the characteristics of being in Melbourne, e.g. for certain administrative purposes such as postal services, but also have some of the characteristics of being outside Melbourne, e.g. for rating purposes. That is, just as the platypus fails to fit eighteenth century zoological classificatory schemes, there is a failure of those locations on the outskirts of Melbourne to fit the classificatory scheme which consists of just the two predicates 'in Melbourne' and 'outside Melbourne', and this failure in turn gives rise to the indeterminacy. Although there is more to it than that. Part of the pressure towards saying Melbourne's area is indeterminate, rather than simply taking each proposition of the form 'Melbourne has area x' as false, comes from the additional taken-for-granted fact that Melbourne is a spatial entity—spread out on a surface—and in particular that we can place determinate limits on 'the area of Melbourne', e.g. that it is more than ten and less than a million square miles in area.

Which of these three distinct traditional sources of indeterminacy that we have discussed fits the quantum theoretical cases? Consider an electron not in an eigenfunction of position, and which therefore, according to the orthodox interpretation, has an indeterminate position. We will take the electron itself to be a determinate entity, like Melbourne and unlike the group of yellow objects on the table. (It is only when it comes to multiple quantum systems, which we will not be considering here,[39] that questions about the determinacy of electrons as such can be raised, by raising difficulties about distinguishing between electrons.)

Can we blame the electron's indeterminacy of position on the indeterminacy of the property of having a particular spatial configuration? It would seem that we cannot for the following reason. In some respects it is as if the electron had one particular position, within some range of positions, i.e. were its position measured it would register at a

---

[39] See Feynman (1965), ch. 2.

single point, whereas in other respects it is as if it were a wave dispersed over some range of positions, i.e. it would exhibit interference effects in a double-slit situation. But the respects in which it behaves as if it had one particular position are not sufficient to entail that it actually has one particular position, and this is also true of the respects in which it behaves as if it were disperse.

Thus we do not have a case of conceptual overdetermination analogous to the case of indeterminate gender discussed above: it is not the case that the electron satisfies *both* the criteria for having a unique position and for being spatially disperse. Nor it seems do we have a case of conceptual underdetermination. The respects in which the electron behaves as if it were spatially disperse are sufficient to say that it does not have one particular position like a particle (so says Bohr), and the respects in which it behaves as if it had a unique position are sufficient to say that it is not spatially disperse like a wave (says Bohr). Thus the criteria for whether a system is spatially disperse and the criteria for whether it has a unique location tell us that *neither* of these situations applies (unlike the case where spatial configuration is simply conceptually underdetermined).

So why do we want to say that the electron's position is indeterminate? I suggest that the electron's indeterminacy of position arises because not only does (a) the electron satisfy some but not all criteria for being spatially disperse but also (b) it satisfies some but not all criteria for being located at a single point, but nevertheless (c) the electron satisfies all the criteria for being a spatial object, situated within some spatial volume, e.g. the universe at large. More particularly if our electron has a state-function which is non-zero over every point in some volume V and only within V then we say that (c)' it is located within V; and we do so because all position measurements will then register it to be located within V with certainty. And it is this triad of statements (a), (b), and (c) which constitutes the grounds for asserting the indeterminacy of the electron's position; and in particular (c)' tells us the bounds of this indeterminacy, i.e. that the electron's position is only indeterminate to within V, and more particularly that the electron is determinately not outside of V.

We thus see the sense in which the concept of having a particular position may be said to 'fit imprecisely' those electrons which have an indeterminate position: it is because the electrons may be said to be somewhere, but there is no precise spatial configuration which they may be said to have. Moreover we see that this case of indeterminacy is precisely parallel to the Melbourne case. In particular it is the failure of the conceptual grid consisting of wave and particle concepts—and more basically the failure of the conceptual grid which distinguishes spatially extended from point objects—to fit electrons, despite the electrons being spatial entities (located within some volume) which gives rise to their indeterminacy of position. And this failure in turn arises because while electrons are spatial objects they only satisfy some, but not all, of the criteria for being extended and some, but not all, of the criteria for being at a point. Just as in the case of Melbourne's indeterminate area certain locations satisfy some, but not all, of the criteria for being inside Melbourne but also satisfy some, but not all, of the criteria for being outside Melbourne.

A similar account of the indeterminacy of other physical quantities can be given. Q is said to be indeterminate in value in S at t if all the criteria are satisfied for Q having *some* value in S at t, e.g. a Q-measurement would reveal some value at t, but there is (are) no particular value(s) for which S at t satisfies all the criteria for Q having that (those) value(s) because of the existence of relevant interference effects. In such a case we will say not just 'Q has an indeterminate value in S at t', but we will also preserve the claim that Q has *some* value in S at t, in just the same way that we said of the electron which had an indeterminate position that it was nevertheless located somewhere (within V say).

There is here a further question of how the notion of indeterminacy is to be treated formally, in particular what semantic structure we are to erect, so that the statement 'Q has an indeterminate value' is consistent with 'Q has some value'. (I take the qualification 'in S at t' for granted.) I shall discuss this question only briefly here, since the issue of 'quantum logic' is beyond the scope of this book;[40] and, in any case, nothing in what follows depends critically on how one answers this question.

One strategy which can be followed to answer this last question is to take the statement that Q has an indeterminate value to imply that 'Q has value q' is not true for all possible q. The problem then is how to reconcile this with an assertion of Q having some value, i.e. with the truth of 'Q has some value' which I take to be equivalent to the truth of the disjunction of all statements of the form 'Q has value q' for all the possible values q (in the case that the domain of possible values is

---

[40] For a discussion see van Frassen (1974) for example.

countable). One way to do this is by extending the classical set of truth-values {true, false} to include a third truth-value—indeterminately true or false—and letting 'Q has value q' have this third truth-value iff it is indeterminate whether Q has value q; and then extending the truth table for disjunction by allowing that a disjunction may be true even though each of its disjuncts is indeterminate in truth-value and hence, in a broad sense, not true. In particular we have 'Q has q_1 or Q has q_2 or . . . ' true iff S at t is in a superposition of eigenstates of Q for values q_1, q_2 . . . or a mixture of such states. This allows 'Q has q_i' to be not true for all i, while the disjunction of all such sentences is true (viz. when the system is in a superposition of eigenstates of Q for each of the q_i).

By allowing this extension of its truth table disjunction does of course cease to be truth-functional, i.e. not all cases of disjunctions of statements with indeterminate truth-values will have the truth-value 'true', e.g. they may have indeterminate truth-value, and hence a truth-value 'not true'. The question of the detailed semantics to adopt here will not be addressed, although I point out that the super-valuational semantics discussed in K. Fine (1975) would seem appropriate. Note however that by construing disjunction in this way the rule of or-elimination fails, a point to which we will return later.

What I am suggesting here (unlike Putnam's suggestion for a non-distributive quantum logic) is only an *extension* of classical logic—not a *rival* to it.[41] In particular we see that the suggestion here is not as radical as Putnam's suggestion to reject the classical truth table for 'or' by allowing a valuation for which 'false or false = true'. What I have suggested merely extends the truth table for 'or' by putting in extra lines. This difference is manifested in the case of the double-slit experiment by Putnam taking it to be false that the electron passes through $slit_1$ and false that it passes through $slit_2$, but maintaining nevertheless that it either passes through one slit or the other; whereas I say that it is merely indeterminate which slit the electron passes through, although I agree with Putnam that it does pass through one or other of the slits.

In sum then the notion of indeterminacy which I am suggesting here has the following logical features. To say that Q is indeterminate in value in S at t does not mean that 'Q has q in S at t' is false for all q: it just means that 'Q has q in S at t' is indeterminate in truth-value for

---

[41] See Haack (1974) pp. 8 ff for this distinction.

certain q, and false for other q (viz. those q for which Q does not have value q in S at t). And this means that 'Q has q in S at t' is not true for each q given that we take 'indeterminately true or false' to mean some third truth-value (neither true nor false). Nevertheless to say that the value of Q in S at t is indeterminate allows one to say 'Q has some value in S at t' is true. (In particular this will be so if one adopts the super-valuational semantics for indeterminate predicates referred to above.) On this view in the double-slit experiment it is true to say that the electron passes through S_1 or passes through S_2 but it is indeterminate which slit it passes through.

Finally in this discussion of indeterminacy, let me dissociate myself from a particular view of the nature of indeterminacy which Bohr's equation of indeterminacy with imprecision of fit might suggest. Suppose we agree that to say some system has a property indeterminately is to say that the objects which exist in the world do not so to speak slot into the spaces which are defined by the conceptual net in terms of which we are representing the world. This metaphor of the conceptual net then generates the notion of imprecise fit, i.e. the term 'fit' metaphorically slips between 'satisfaction' as applied to concepts and 'fit' as in objects fitting into some slot. On this view the existence of indeterminacies emerges as a flaw, to be rectified by the adoption of a more appropriate conceptual net. And at first sight it seems that this appropriate conceptual net is supplied by QT. The language of state-functions provides a perfectly precise means of describing q-systems, even to the extent of providing (as we shall see shortly) explications of the notion of their degree of imperfection of fit. In the same way, one may argue, the vague terminology of colours can be replaced by the precise terminology of the radiation reflection characteristics of surfaces. In Dummettian terms (see footnote 38 above) one would say that the quantum theoretic indeterminacy is not intrinsic, because precisification of the vague terminology is possible (unless of course one followed Bohr in taking the classical description as somehow unavoidable).

But an indeterminate position can be seen in a quite different light, as being a new sort of property, different from that of having any particular position(s) precisely, but theoretically respectable in its own right. And it is that view which I shall be taking here. It is important to note that the introduction of such new properties does not reverse the programme of eliminating 'indeterminate' in the broad sense of 'vague' or 'imprecise' terminology from the language of physics, which characterized much of classical physics.[42] What has happened is that the concept of indeterminacy has itself been precisified and incorporated into theoretical discourse. Thus the programme of eliminating indeterminate (in the broad sense) terminology has actually been furthered by the paradoxical device of including the term 'indeterminate' within QT.

7. DEGREES OF INDETERMINACY

I shall now complete my account of the notion of indeterminacy within the new QT by introducing the idea of *degrees* of indeterminacy. I shall do this in the context of a more general summary of the orthodox (or Copenhagen) interpretation of QT. The version of the orthodox interpretation of QT which I shall give here can be found in one form or other in many texts on QT, e.g. Schiff (1955) and Feynman (1965), and is in particular based on the version promulgated by Pauli in his influential *Handbuch der Physik* article of 1928.[43]

According to the orthodox interpretation of QT there is for every *q*-system and time (during the lifetime of the system) a set of physical quantities to which we can assign determinate numerical values. (What this set is may vary for different systems and times.) Moreover such physical quantities as appear in QT—the '*q*-quantities'—also belong to the set of physical quantities which appear in classical mechanics. But in QT (unlike classical theories) we allow that some physical quantities may have indeterminate values. Indeed it is postulated that there are pairs of physical quantities for which if one of the pair has a determinate value at some time then the other member of the pair must have an indeterminate value at that time. The members of such a pair of physical quantities are said to be incompatible, and the pair is said to constitute a conjugate pair.[44] For example the momentum and position of a given *q*-system in a

---

[42] See Carnap (1966), pt. 2 for example. For Carnap this programme took the form of the precisification of vague qualitative terminology by precise quantified terminology. Thus the relation of being hotter is explicated by that of having a greater temperature.

[43] See Pauli (1933).

[44] In more modern terminology being a conjugate pair means having no eigenvectors in common, a special case of which is having a constant but non-zero commutator.

particular direction form such a conjugate pair: if the momentum is determinate then the position is not and vice versa. The existence of such conjugate pairs means that indeterminacies are inescapable in QT. If we prepare a system in such a way that one of the q-quantities in a conjugate pair, say Q_1, has a particular value precisely at some time then the q-quantities incompatible with Q_1 will be indeterminate in value at that time. Of course not all pairs of q-quantities are conjugate. Indeed any q-quantity will belong to a whole set of q-quantities which are all compatible (i.e. not incompatible) with one another. Any such set of mutually compatible q-quantities is said to be complete if there are no q-quantities outside the set which are compatible with all members inside the set.

To stipulate the values taken at a given time by all of the q-quantities for a q-system S is impossible if there are incompatible q-quantities for that system (since this means that if some of the q-quantities have determinate values then others do not). We can however stipulate determinate values for the q-quantities belonging to one of the complete sets for S at a given time (which complete set it is will depend on the state of S at the time). This information is then said to constitute a picture of S at t (in explication of Bohr's notion of a picture).[45] That is a picture of S at t is a statement of the values at t for each of some complete set of q-quantities. There are however many different possible complete sets although only one of them can be the set which in fact provides a picture of S at t. And which one that is will depend on what state S is in at t. Indeed a (non-degenerate) q-quantity Q has the value q determinately in S at t iff the state of S at t is represented by one of a set of 'eigenvectors' of Q, viz. the eigenvector of Q for value q (what 'eigenvectors' are will be discussed in the next chapter). (Note however that this principle only holds if we consider pure states, which is what I am implicitly doing in this chapter.)

But now for a complication. There will be times t at which the q-system S will be fitted by no pictures at all, e.g. both momentum components and position coordinates for S will be indeterminate at t. Nevertheless we can define a sense in which both the wave and particle pictures will fit *approximately* S at t, as follows. Suppose that a measurement of three momentum components for S at t would yield various measured values with varying probabilities. Suppose that the

---

[45] Although note that once we allow in mixed states we may be able to stipulate values for more physical quantities for S at t than are contained in one complete set.

probability of the result of the measurement of the x_1-component of momentum differing from the value p_1 say, by more than an amount Δp_1 is negligible (by some agreed standard of what is to count as negligible).[46] Then we say that it is approximately determinate that S at t has a momentum component in the x_1 direction of value p_1, where the degree of approximation here is defined as Δp_1. In other words the closer it is to certainty that a measurement of some physical quantity will yield a particular value the better is the approximation in saying that the relevant quantity has that value (determinately), and in that sense the less indeterminate is the physical quantity. Now let $\Delta p_1, \Delta p_2, \Delta p_3$ be the degrees of indeterminacy (as defined above) for the three momentum components and $\Delta q_1, \Delta q_2, \Delta q_3$ be the corresponding degrees of indeterminacy for the three position coordinates. Then both the particle and wave pictures are said to fit approximately, with the relevant degrees of approximation being the upper bounds on $\Delta p_1, \Delta p_2, \Delta p_3$ and $\Delta q_1, \Delta q_2, \Delta q_3$ respectively. QT implies that the degrees of indeterminacy so defined satisfy the Heisenberg indeterminacy principle:[47]

$\Delta p_i \Delta q_i \geq \hbar$ for $i = 1, 2, 3$.

We now see that the orthodox interpretation, which I have just sketched, does indeed differ considerably from the views which I attributed to Bohr above with respect to the role assigned to the notion of degree of indeterminacy. For Bohr the primary phenomenon in QT in terms of which all else is explained is that of wave–particle duality, i.e. the complementarity of the wave and particle pictures, which means that there are reciprocally related restrictions on how the wave and particle pictures can be fitted to a q-system: if the wave picture fits exactly then the particle picture does not, and vice versa. Moreover Bohr also defined a degree of imprecision with which pictures can be fitted to q-systems (although in a somewhat different way from the above), and claimed that these were reciprocally related q.v.

(The degree of imprecision with which one picture fits S at t) × (The degree of imprecision with which the complementary picture fits S at t) $\geq h$.

---

[46] In QT we traditionally let Δp_1 be the standard deviation of the probability distribution over the various possible results of measurement.

[47] See Heisenberg's derivation in (1930) pp. 15–19.

And he demonstrated this relation in a well-known series of thought experiments.[48] On the other hand the orthodox view (Pauli's) takes the indeterminacy relations to be directly deducible from the laws of QT by defining indeterminacy in terms of spreads of measured values (see footnote 48). On the latter view the wave–particle duality is seen as merely a colourful way of bringing out the consequences of these theoretically derived indeterminacy relations.

[48] See Heisenberg (1933) pp. 21 ff., or Bohr's debate with Einstein in Bohr (1949).

2

The State-vector and Probabilities

Quantitative concepts; phase-space; abstract spaces; vectors; probabilities and propensities; the Born interpretation; degenerate physical quantities; state-vectors as properties of single systems.

1. PHASE-SPACE

One of the features of modern physics is the extensive use of quantitative concepts to enrich qualitative concepts. For example a comparative qualitative judgement that A is hotter than B will be redescribed quantitatively by saying that the temperature value of A is greater than that of B. This quantitative language has much greater expressive power than the qualitative language it replaces. For example it enables sense to be made of such claims as 'A is as much hotter than B as B is hotter than C'. For our purposes, however, the important feature of quantitative language is that it enables the development of the idea of a possible state.

For example consider the way we quantify the qualitative concept of being 'hotter than'. We do this by setting up a temperature scale. This can be seen as a mapping of bodies in states of thermal equilibrium onto a subset of the real numbers in such a way that A is hotter than B iff the temperature of $A >$ temperature of B. The number onto which a body is mapped by this scale is said to be its temperature according to that scale. It is characteristic of the Celsius scale for example that freezing water (at one atmosphere pressure) is assigned the temperature value 0, and boiling water is assigned the temperature value 100. Other bodies are then assigned temperature values on the Celsius scale by fixing equality of temperature differences. (One way of doing this is to say that if a given mass of ideal gas changes its volume by the same amount in changing its temperature from T_1 to T_2, *ceteris paribus*, as it does in changing temperature from T_3 to T_4, *ceteris paribus*, then those temperature intervals have the same magnitude).[1] The subset of real numbers which is the range of the Celsius scale is then taken to constitute the set of 'possible values for temperature'. It is a continuous sub-interval of the reals from -273 to plus infinity.[2]

We can describe what has been done in constructing the Celsius scale as follows. The properties of certain bodies, namely the 'coldness' of freezing water and the 'hotness' of boiling water, are represented within a network of numbers which form a continuous sub-interval of the real numbers. Moreover a procedure is specified which represents the hotness (or coldness) of other bodies by points in that network, although we do not know in advance of putting the procedure into operation which points in the network will be 'occupied', i.e. which numbers will be the temperature values of real bodies at particular times. The numbers on the scale are taken as possible temperature values and can be seen as representing possible states of hotness, in the way that spatial coordinates represent possible locations in physical space say. Thus in the process of quantifying the qualitative concept of hotness we enrich the concept: we provide a notion of 'degrees' of hotness—a spectrum of possible states of hotness—which can be linearly ordered and represented numerically on a scale.

Moreover we see that the value of the temperature of a body as it changes over time may be regarded as 'moving up and down' the numerical range of the temperature scale, and hence the value of temperature of the body may be seen as moving through the space of possible temperature values defined by that scale, where the term 'movement through the space' is of course to be interpreted metaphorically. This metaphor has as its point of origin the movement of bodies through what we will call 'physical space', the space in which all bodies are located. It must be pointed out however—and this is a

---

[1] Note that there is a question here of whether there is some independent notion of equality of temperature differences which must be reflected in all temperature assignments. If there is then the centigrade scale—or indeed any temperature scale—is determined by just two fixed points.

[2] It is a law of thermodynamics that there is an 'absolute zero', i.e. a lower bound on temperatures.

point we shall emphasize later in the context of QT—that this metaphor, like any metaphor, is misleading at certain points. For a start it is the body itself—not its location—which moves through physical space. Moreover physical space is not like the set of possible temperatures in that physical space is not a set at all, and is certainly not the set of possible locations for bodies. If anything it is to be seen as the mereological *sum* of such possible locations, or more correctly as the sum of space points.

Various of the properties of a body, e.g. its mass, charge, position, etc. can be seen as the results of quantifying qualitative relations as in the case of temperature. Indeed for the point particle of Newtonian mechanics all of its fundamental mechanical properties at a particular time—its mass, velocity, position—can be seen in this way. In particular each of these properties is represented by some number/set of numbers (as the temperature of a body is), e.g. the value of its mass, the values of its velocity components, its position coordinates, etc.

The 'state' of a Newtonian point particle S at time t—what we may call 'its Newtonian state'—can then be defined simply as the conjunction of all these fundamental properties for S at t. This 'state' can be seen as what is responsible for all the mechanical effects which would (under suitable conditions) be manifested by S at t, e.g. how S at t would collide with other particles, etc. The state of S at t can be represented by six numbers which are its position coordinates and the values of its velocity components at t in certain specified directions.[3]

We can then formally construct a set (or 'space') of 'possible state representations' for S. This is simply the set of all possible combinations of values for these six fundamental physical quantities, and is the set of (ordered) sextuples of numbers of form $(x_1, x_2, x_3, v_1, v_2, v_3)$, where x_1 ranges over all possible position coordinates in the first direction, v_1 ranges over all possible values for the first velocity component, etc. This space of possible state representations—called 'the phase-space for S'—can then be (metaphorically) regarded as the space through which the state of S (or really the *representation* of the state of S) moves as the mechanical properties of S change over time.

---

[3] The mass is not included in the state only because it is seen as constant in time, so if it is specified at just one time then it can be set aside. It is also conventional not to include the angular momentum of the point particle in its state, or at least to treat it separately—as an aspect of the rotational state of the particle.

The sextuple of numbers which represents the state of S at time t is called 'the representative point for S at t'.[4]

Formally we can also define a 'space of possible states' as a set of elements which are represented by the elements in this space of 'possible state representations'. Or more directly we can define 'possible states' as possible states of affairs which are described by certain logical constructions (indeed conjunctions) or propositions which assign possible values to physical quantities. Thus the possible state $(x_1, x_2, x_3, v_1, v_2, v_3)$ for S at t is the possible state of affairs of S at t having value x_1 for its first position coordinate, *and* having value x_2 for its second position coordinate, etc. Each element in the phase-space is mapped onto one possible state which it is said to represent, and vice versa. In the same way the space of 'possible temperature values' on the centigrade scale was used above to define a space of possible states of hotness, each distinct value on the scale representing one and only one distinct possible state of hotness; or, to push the metaphor back to its root, in the same way as the 'space' (here meaning the set) of possible space-coordinates represents that set of possible locations which together make up physical space. Derivatively we can then say that the state of S at t moves through the space of its possible states over time.

This notion of a space of possible states appears rather trivial in the context of classical mechanics, because of the possibility of describing possible states simply by conjunctions of propositions assigning possible values to physical quantities. And correspondingly the space of possible state representations is simply a 'gluing together' (formally a direct product) of the spaces of possible values for each of the relevant physical quantities. In QT however, as we shall see, this is not the case; and hence the notions of a state and of a phase-space become correspondingly more interesting. In particular the relation between a state assignment and the values taken by physical quantities no longer takes the simple form which it takes in the classical context. That is in the classical context each of the N-tuple of numbers representing the state of a system is the value of a distinct physical quantity; but in QT, although states may be represented by N-tuples of numbers (representing vectors), it is *not* simply the case that each such number is the value taken by a physical quantity.

---

[4] This notion of a phase-space and a representative point moving through it was first developed in Gibbs's classical statistical thermodynamics.

What I have said so far has not considered whether there really is a property of having a particular temperature, or of having a particular mass, or spatial location. One view is that the whole point of introducing the concept of temperature is to eliminate the 'un-scientific' notion of hotness. Thus instead of saying that the tempera-ture value of X at t represents the 'hotness' of X at t we can—and according to this view should—take it as representing the *temperature* of X at t, where the latter is now conceived of as a property in its own right explicative of the pre-theoretical concept of hotness.

This point of view contrasts with Carnap's (1966) views. For Carnap the 'real properties' are precisely the qualitative relations of being more or less hot; and the quantitative properties of having a particular temperature provide a convenient, but conventional, way of representing the former properties. In particular one takes statements like:

(S) X at time t has a temperature of 50°Celsius

to have a simple subject–predicate form, instead of the form:

$(S)_1$ The temperature of X at $t = 50°$Celsius,

where '50°Celsius' names a particular property. Or at most one might postulate a function 'temperature', or better 'temperature in degrees Celsius', which maps bodies and times onto numbers, so that (S) takes the form:

$(S)_2$ $T(X, t) = 50$,

where T is the relevant function 'temperature in degrees Celsius'. This latter analysis preserves the extensionalist programme put forward by nominalists about properties. (Note that on this last view to talk of the value of the temperature of X at t is redundant: it is no more or less than the temperature of X at t. This contrasts with the view that the temperature of X at t is a *property* which is represented numerically by the 'value of the temperature of X at t'.)

I point out these various possible views here, in order to show that the ontological question of whether the properties of having particu-lar temperatures (or mass, or spatial location) exist is not to be settled just by the theoretical formalism. The naïve idea that we can simply read off our ontological commitment from our best theories is just that—naïve. Indeed the example of temperature suggests that, if anything, the contrary is the case; we bring to our theories an ontological commitment which we then use to divide the real from the conventional in our theoretical representations of the world. This question of the interaction between the metaphysical, the real, and the theoretical will be alluded to later in this chapter.

2. ABSTRACT SPACES

Complementary to the notion of a space of possible states (and a phase-space) is the notion of an abstract space. An abstract space can be seen as an abstraction from what we may call 'position-space'. Position-space is the set of possible positions in physical space, i.e. position-space is the set of space points. So position-space can be seen as the set of entities which when summed together (in the mereological sense) make up physical space. The notion of an abstract space then arises when we start to consider the structure of position-space as defined by the structure of the corresponding physical space, in the following way.

The three-dimensional Euclidean nature of physical space lies in the fact that with every point P in physical space we can associate three numbers called 'the Cartesian coordinates of P' for which the distance $S(P, P')$ between two points P and P' in physical space is taken to be given by the Pythagorean relation:

(P) $S(P, P')^2 = [X_1(P) - X_1(P')]^2 + [X_2(P) - X_2(P')]^2 + [X_3(P) - X_3(P')]^2,$

which we abbreviate to:

$$\sum_{i=1}^{3} [X_i(P) - X_i(P')]^2.$$

And we can then derivatively say that position-space (construed as the *set* of points in physical space) is three-dimensional Euclidean in virtue of the fact that for every pair of points (possible positions) P, P' which belong to the space the equation (P) just given holds.

The previous account of the Euclidean nature of position-space leads naturally to the idea of an *abstract* Euclidean space. An abstract space is regarded as a set of 'elements' for which the identity of the elements is not specified except in so far as they have certain structural properties or relations. For example the relevant structural properties may simply be that the elements altogether can be mapped one-to-one onto the set of triples of real numbers and that there is a binary function S defined on the elements for which for any elements P, P'

$$S(P, P') = \sum_{i=1}^{3} [X_i(P) - X_i(P')]^2,$$

where $(X_1(P), X_2(P), X_3(P))$ is the triple of numbers onto which P is mapped.

The abstract space so defined is said to be Euclidean for obvious reasons. Note however that it is misleading to say of such an abstract space that it is purely mathematical, as if its elements were somehow purely mathematical in nature. The nature of its elements is not specified at all. Indeed the only sense in which the elements of abstract space may be said to exist is the sense in which the elements of some *instance* (or realization) of an abstract space may be said to exist, where by an 'instance of some abstract space' I mean any space, i.e. set, of entities which has the same structural properties as those which define that abstract space. Those entities in turn may be purely mathematical, e.g. numbers, or they may be physical (points of physical space), and hence be said to exist or not depending on one's ontological commitment.

What is the point of studying such abstract spaces? It is because in studying them we can discover which properties of a particular set of objects (say the points in position-space) follow merely from those objects having certain structural attributes, for example a Euclidean distance function. For a similar reason we study abstract set theory, i.e. in order to establish relations between particular sets which follow not from the nature of the individual members of the sets but from certain broader 'set theoretic' features of the sets. For example the set of one-cent pieces in my pocket is a proper subset of the union of the sets of one-cent and two-cent pieces in my pocket, but this fact has nothing to do with what the members of those particular sets are, but merely with the relevant sets being non-empty and bounded, i.e. with purely set theoretic properties.

Yet other abstract spaces are produced by abstracting further from three-dimensional Euclidean position-space. For example an abstract N-dimensional Euclidean space (of dimension N which may be greater or less than 3) is an abstract space in which each element P is represented by N numerical coordinates $(X_1(P), X_2(P), \ldots, X_N(P))$, and the distance $S(P, P')$ between P and P' is given by the generalized

Pythagorean relation:

$$S(P, P')^2 = [X_1(P) - X_1(P')]^2 + [X_1(P) - X_2(P')]^2 + \ldots$$
$$+ [X_N(P) - X_N(P')]^2,$$

which is abbreviated to:

$$\sum_{i=1}^{N} [X_i(P) - X_i(P')]^2.$$

For many situations however the structure exhibited even by an abstract N-dimensional Euclidean space is too rich. Certain elements of structure present in an abstract N-dimensional Euclidean space will be relevant to a situation but not others. For example we may have a space of elements in which no physically significant set of coordinates is associated with a particular point, but there is nevertheless a significant 'distance' function which can be defined on the set of pairs of points. That is there is a significant function S satisfying the three relations:

(i) $S(P, P') = 0$ if $P = P'$

(ii) $S(P, P') \geq 0$

(iii) $S(P, P'') \leq S(P, P') + S(P', P'')$.

Any such function is called 'a metric', and an abstract space on which such a function is defined is called 'an abstract metric space'.

Much of modern analysis is founded by abstracting structural elements from position-space in just this way (the theory of abstract metric spaces, topology, differential geometry, abstract vector space-theory, etc.). Indeed in general relativity we apply one of these more general frameworks (that of differential geometry) back onto position-space itself. That is in general relativity position-space is taken to be an instance of an abstract differential space which is more general than abstract three-dimensional Euclidean space. Thus position-space, and hence physical space, turns out not to have been as simple as we once thought, although the structures which were abstracted from our original over-simplified ideas of its nature turn out to be adequate to express what we now take to be its true nature. Thus we have a feedback loop between mathematics and physics, with mathematics feeding off an initial physical theory, abstracting structures from it in order to develop a new mathematical framework, and this new framework being applied back to frame a new physical theory which replaces the initial one.

The two developments in the notion of a space which are discussed in sections 1 and 2, viz. the development of the notion of a space of possible states and the development of the notion of an abstract space, go hand in hand. As more sophisticated abstract spaces become available the possibility is opened up of more subtle physical theories in which the spaces of possible states become instances of the more sophisticated abstract spaces. And it is to a discussion of just such a development in the context of QT that we shall now turn.

3. ABSTRACT VECTOR SPACE

In QT as in other theoretical contexts the states of systems are seen as categorical bases for certain ranges of dispositions. They are conceived of as physical properties of q-systems responsible for certain (dispositionally construed) quantum effects, in just the way that the fragility of a piece of glass is taken as responsible for its breaking (under stress), or the 'states of hotness' (temperatures) of bodies are put forward as causes of thermal effects, or the Newtonian states of point particles are taken as responsible for certain mechanical effects. But whereas the states of hotness and the Newtonian states are simple logical constructs out of certain independently given properties of bodies, viz. the properties of physical quantities taking particular values, this is not so for the states of QT. In QT we define the concept of a state *ab initio* (take it as a primitive) and then relate it to the values taken by physical quantities, or at least to the probabilities of those physical quantities being measured to have certain values, via the laws of the theory. Moreover although it is—under certain conditions to be discussed—possible to eliminate talk of states from QT in favour of talk about the probabilities of measuring physical quantities to take particular possible values, this elimination hugely complicates the theory, and is in any case (as we shall see) not always possible. So even if we broaden the notion of physical quantities to include those probabilities, the connection between states and values of physical quantities in QT differs significantly from the corresponding relation in the classical context.

These points will now be illustrated by discussing the representation of states in QT. In QT the space of possible states is an instance of a particular sort of abstract space, viz. an abstract vector space. In order to explain what an abstract vector space is we shall first need to say what a vector in physical space is. A vector in physical space—a 'physical vector'—can be seen as an ordered pair consisting of a length (or magnitude) and a direction in physical space. These physical vectors have certain properties in addition to their length and direction, e.g. certain relational properties such as a certain angular separation from other physical vectors. (The angular separation between two physical vectors is simply defined as the angle between their respective directions.) Moreover any physical vector of length L and direction d can be uniquely associated with a coordinate triple (x_1, x_2, x_3), which is simply the set of coordinates of that point in physical space which is reached by moving out a distance L from the origin along the direction d. Indeed, in some texts, the physical vector (L, d) is simply defined to be the directed line segment connecting the origin with the point (x_1, x_2, x_3); or (more perspicuously) is defined to be the equivalence class of such line segments, all with length L and direction d. A physical vector (L, d) may also be defined as an operation on physical space, viz. that transformation which displaces points a distance L in direction d. But the latter definitions, although possible ways to define physical vectors, are not as convenient for our purposes.

We now construct an abstract vector space—a space (or set) of elements which are defined only in so far as they share certain of the 'structural' properties of physical vectors. The properties of physical vectors from which we choose the properties to define an abstract vector space are the following. Each physical vector f can be uniquely associated with a triple of numbers (F_1, F_2, F_3), which are the coordinates of the end-point of the vector f when its other end-point is made to coincide with the origin (the point which is arbitrarily assigned coordinates $(0, 0, 0)$). Moreover it is easy to prove that $|f|$, the length of f, is given by:

$$|f|^2 = F_1^2 + F_2^2 + F_3^2.$$

And we can define a binary operation on pairs of vectors, that of taking the scalar product, by the equation:

(f, g), the scalar product of f with g, $= F_1 G_1 + F_2 G_2 + F_3 G_3$.

So defined it is easy to show that (f, g) is a measure of the degree of closeness of f and g in the sense that, for f and g of given length, (f, g)

has a maximum value when $f = g$ and a minimum value of 0 when f is at right angles to g.

We now define an abstract vector space as any set with elements $f, g, \ldots$, which can be mapped one-to-one onto N-tuples of real numbers, with f mapped onto $(F_1, F_2, \ldots, F_N)$ say. We define $(f, g) = F_1 G_1 + F_2 G_2 + \ldots$ as above, but allow the number of coordinates representing each abstract vector to be more than 3 although the same for all vectors in a given space. If we consider an abstract vector space in which an infinite coordinate set $(F_1, F_2, \ldots, F_i, \ldots)$ represents each abstract vector f then it does of course become a non-trivial question whether $|f|$, the limit of $\Sigma_i F_i^2$, exists, and whether (f, g), the limit of $\Sigma_i F_i G_i$ exists, for particular f, g. But it is definitive of abstract vector spaces that such (f, g) exist for any f, g in the space, and hence of course that $|f|$ exists for any f, being just $(f, f)^{1/2}$. A further generalization to be discussed later is to allow the various F_i to be complex numbers rather than real numbers. Such abstract spaces (with certain minor modifications to be discussed later) are called 'abstract Hilbert spaces'.

Now we can understand the claim that the set of possible states provided by QT for the system S is an instance of an abstract vector space, indeed an instance of an abstract Hilbert space $\mathscr{H}(S)$. It means *inter alia* that the set of possible states can be mapped one-to-one onto the set of N-tuples of real numbers (for some N), and that there is an appropriate scalar product operation definable on pairs of such possible states. (In the same way we say of the temperature scale that it forms a linear ordering, where this means that it shares certain structural features with lines in physical space.) The N-tuples of numbers onto which possible states for QT are mapped are said to represent those states, in the same way that temperature values represent states of hotness. For simplicity and to conform with normal usage we will abbreviate the claim 'the set of states of S is an instance of an abstract vector (Hilbert) space' to 'the set of states of S is a vector (Hilbert) space', although one must of course distinguish this last claim from the false claim that the states of S at t are physical vectors. Rather the states of S at t are members of an instance of an abstract vector space. This last distinction becomes clearer when we ask in what sense we can say that a state in QT has a length (*sic*). We can now answer this question in terms of the abstraction argument developed above. In the first instance we talk of physical vectors having a length (a length as defined on physical space—construed as

Euclidean).[5] We then 'go up a level', to talk of abstract vectors defined as vehicles for certain generalized properties of physical vectors, including the property of having length. And we can then derivatively assign a 'length' to any concrete instances of these abstract vectors, including the possible states of QT.

My terminology here is slightly at variance with more traditional usage according to which the state of S at t is either said to be represented by a vector or is even taken as a vector itself. I reject these usages because it leaves puzzling just what is meant by 'vector' (an element of an abstract vector space, a physical vector, an element of an instance of an abstract vector space, or perhaps the N-tuple of numbers associated with a vector). Nevertheless in what follows I shall often lapse into this more traditional usage if only to expound some traditional view in a recognizable way.

We can now see that the Hilbert space $\mathscr{H}(S)$ associated with a q-system S is not the space in which S is located or moves, at least not in the ordinary sense of those terms (i.e. it is not physical space). Nor does there seem to be any reason to stretch ordinary usage here. Rather it is a space, i.e. set, associated with S at t, the members of which are the possible states of S at various times. One must therefore expunge from one's thoughts those popularizations of QT which say that electrons move in some higher-dimensional (indeed infinite-dimensional and complex) space called 'Hilbert space' rather than in ordinary (finite-dimensional and real) physical space.[6] There is a grain of truth behind these popularizations, viz. that there are difficulties in talking about the location of electrons because of their indeterminacy in position, but this is not because electrons are 'really' located in some mysterious Hilbert space. The Hilbert space for an electron does not pretend to provide a locus for the electron; it is simply a stock of possible electron states. Indeed the only sense in which electrons move in Hilbert space is the metaphorical sense discussed above in which the state of the electron may be said to move through the space of its possible states, in the same sense that the

---

[5] Note that it is irrelevant here that physical space is not in fact Euclidean, because physical space is being used only in a heuristic role, as a base from which to abstract a structure.

[6] See for example Zukav (1980) p. 99: 'The quantum leap is also a leap from a reality with a theoretically infinite number of dimensions into a reality which has only three. This is because the wave function of the observed system, before it is observed, proliferates in many mathematical dimensions.'

temperature of a body may be said to move up and down the temperature scale.

4. PROBABILITIES

So far we have not said how the states of q-systems are connected with 'observables' ('phenomena' is perhaps a better term here).[7] Classically this connection is particularly simple. The state of a classical point particle is represented (within phase-space) by a set of numbers which are the values of its three velocity components and its three positional coordinates, and it is the values of these physical quantities which are the 'observables'. In QT states are also represented by N-tuples of numbers but at one remove so to speak. States are identified with abstract vectors, and these vectors are in turn represented by, i.e. mapped one-to-one onto, N-tuples of numbers (which are the coordinates of the end-points of the vectors when they are placed at the origin). Thus in a sense states may still be seen as represented by N-tuples of numbers. But it is as we shall see no longer the numbers themselves which are physically significant. Rather it is the angles *qua* scalar products between the vectors which are physically significant, because they are identified with the probabilities of physical quantities taking (or at least being measured to have) certain values. These angles between elements in phase-space have no physical significance at all classically but they are all that is significant in QT.

Thus in QT there are two important differences from classical physics. First it is the angles in the phase-space which are significant, not the coordinates of the points. Second it is probabilities associated with the possible values for physical quantities, rather than the values of the physical quantities, which function as observables. In the rest of this section we shall pursue the second of these points; and in the next section we shall pursue the first.

Historically the first choice made for observables in QT were the various probabilities that some q-quantity Q has its ith value in system S at time t, for various S, t, and i. More particularly Born originally put forward his statistical interpretation to connect $f(S, t)$, the state-vector for S at t, with the probability of S at t being within a small

---

[7] See Hacking (1983) for a critique of the notion of 'observable', as opposed to 'phenomenal'. We will return to this distinction later.

interval dx of the position x, for various x.[8] (It is important to note that these observables chosen by Born are not observational in any sense which implies that they are somehow epistemologically secure or at least more secure than 'theoretical' quantities. This is because the estimation of values for the relevant probabilities involves both the acceptance of a measurement theory for the relevant physical quantities as well as a technique for estimating probabilities from relative frequencies which allows for some finite probability of error.)

This initial choice for the observables in QT was undercut by the Bohr interpretation of QT because according to Bohr a q-quantity Q does not have a value (or at least no determinate value) for S at t, except in very special circumstances. *A fortiori* except in special circumstances the probability of Q having value q_i in S at t is not determinate, or worse is zero, for all i.

The solution to this problem was to change the relevant probabilities in terms of which $f(S, t)$ was to be interpreted, to the probabilities $P(Q, q_i, S, t)$ of *measuring* Q to have value q_i in S at t for various Q, q_i. To a verificationist of course this change was more by way of unpacking what was meant in the first place. After all what else could a verificationist mean by 'the probability of Q having value q_i' than 'the probability of Q being measured to have value q_i'? To a non-verificationist however the change here is of radical significance: from talk about what is to talk about what is measured, or even just what could be measured.

There are serious problems created by this change. Not least there is the problem of interpreting what 'the probability of Q being measured to have value q_i in S at t' means. There are several candidates. It cannot mean the probability that Q is measured to have value q_i in S at t *simpliciter*, because the latter probability can trivially be made 0 by arranging matters so that there is zero probability of measuring Q in S at t (e.g. if measuring apparatuses for Q in S at t are all arranged to explode). Clearly the $P(Q, q_i, S, t)$ of QT are not intended to be made 0 in this fashion. Several candidates for the meaning of $P(Q, q_i, S, t)$ remain:

(A) $P(Q, q_i, S, t)$ is the probability of the subjunctive conditional 'Were Q measured in S at t then the measured value would be q_i'.

---

[8] See Born (1926). Note that it is traditional to refer to the physical quantities themselves, rather than their probabilities, as observables in QT, but this usage is not followed here.

(B) $P(Q, q_i, S, t)$ is the conditional probability that q_i is the measured value of Q in S at t, given that a measurement of Q in S at t takes place.

One could even syncategorematically analyse '$P(Q, q_i, S, t)$' by taking '$P(Q, q_i, S, t) = p$' to mean:

(C)   If Q is measured in S at t, then $P(Q$ is measured to have value q_i in S at $t) = p$,

or:

(D)   Were Q measured in S at t then $P(Q$ is measured to have value q_i in S at $t)$ would $= p$.

(The difference between (C) and (D) is clear if the 'if-then' in (C) is construed truth-functionally: (C) is trivially true if Q is not in fact measured in S at t. On this construal (C) is clearly unsuitable as an analysis of '$P(Q, q_i, S, t) = p$'.)

Physics texts give little if any explicit guidance on which of these options to adopt. Does the practice of physicists gives us guidance? Consider how the statement '$P(Q, q_i, S, t) = p$' is confirmed/disconfirmed experimentally in the context of QT. One tests '$P(Q, q_i, S, t) = p$' by preparing a large number N of systems all in the same state as S at t by following one of the recipes for state preparation supplied by QT,[9] performing a Q-measurement on all of them, and then evaluating the relative frequency with which q_i is the measured value. This relative frequency is then taken as a good estimate of p for N large (with standard statistical estimating methods determining the goodness of the estimate).[10] This estimating procedure provides us with no indication of which of (A) to (D) is correct however. In fact we can generalize here and say that from a pragmatic point of view—a point of view which favours experimental rather than metaphysical concerns—the differences between (A) to (D) are of no significance.

Nevertheless the principles of how to test probability assignments which we have just introduced do provide some insight into the nature of the $P(Q, q_i, S, t)$ even if they do not resolve the particular issue between (A) to (D). In particular they suggest one of two views of the relevant probabilities *qua* objective probabilities. First there is the

---

[9] 'All electrons in the ground state of a hydrogen atom are in the same state' is one such recipe.

[10] The assumption is of course that these state preparations are independent in the relevant sense.

frequentist view which simply defines probabilities to be identical with long-run (perhaps even the infinitely long-run) relative frequencies over ensembles of similar systems. In particular $P(Q, q_i, S, t)$ is identified with the relative frequency with which Q is measured to have value q_i over an ensemble of Q measurements on systems in the same state as S at t. This interpretation has the advantage of relating the latter relative frequency directly to $P(Q, q_i, S, t)$ in such a way as to explain (albeit trivially) why that relative frequency is used to estimate $P(Q, q_i, S, t)$. On the other hand this interpretation has the well-known disadvantage[11] that it is hard to do justice to the idea that the latter probability is a property of the *particular* situation in which S is located at t. In other words the surface grammar of the probability assignment, which assigns the probability to the particular S, t (as arguments of the probability function), is simply ignored.

The second possible interpretation for $P(Q, q_i, S, t)$ is to take it as a 'single-case probability', a relational property of a particular trial (measuring Q in S at t) and a particular type of outcome (measured value q_i). This interpretation (or really partial interpretation—see below) at least has the advantage of doing justice to the surface grammar of $P(Q, q_i, S, t)$; but it has the disadvantage that it leaves puzzling why a long-run relative frequency should be a measure of $P(Q, q_i, S, t)$. The way to explain this puzzle is well known, via the law of large numbers.[12] We take it that $P(Q, q_i, S, t)$ only depends on S at t via $f(S, t)$. Then, by the law of large numbers, it follows that the relative frequency of the measured value of Q being q_i over a reference class of independent measurements of Q on systems at times when those systems are all in the same state as S at t is 'approximately' equal to $P(Q, q_i, S, t)$. (The 'approximation' here is in a statistical sense only, i.e. the probability of there being any particular finite difference between the relative frequency and $P(Q, q_i, S, t)$ tends to zero as the number of members of the reference class tends to infinity.) This result explains why, and in what sense, the relevant long-run relative frequency is a good estimator for $P(Q, q_i, S, t)$. The explanation is less direct—less trivial one might say—than the explanation offered by the frequency interpretation, because the connection between relative frequency and probability which it proposes is less direct. But in retrospect we realize that the connection between relative frequency

---

[11] See Mellor (1971), ch. 3 for a discussion of these issues.

[12] For example see the last section of Lewis (1980).

and probability proposed by the frequency interpretation is too direct: we surely want to allow at least the possibility (albeit an improbable possibility) of long-run frequencies diverging, 'by chance', from the corresponding probabilities. The single-case interpretation has the advantage that it allows precisely that possibility.

In sum then, I take the previous arguments to come down in favour of the single-case interpretation of the $P(Q, q_i, S, t)$. But the latter is still not a full interpretation of $P(Q, q_i, S, t)$. For example although I have taken $P(Q, q_i, S, t)$ as a single-case probability, and hence as a property of, *inter alia*, the particular trial of measuring Q on S at t, I have not said whether it is a property of just S at t, or a property at t of the 'whole experimental set-up' including S. (The latter view is Popper's—see his (1983b) p. 356.) Nor have I said what sort of a property this is. I shall not commit myself on the former point, but on the latter point I shall follow Mellor (1971) in taking the objective single-case probability of an outcome of type E for a certain situation S to be the display of an underlying propensity for E to occur in situation S. In particular I shall take the fact that $P(Q, q_i, S, t)$ takes a certain value—say p—to display the existence of a particular propensity for the measurement of Q in S at t to have the result q_i.

The term 'display' here may seem inappropriate. After all there seems to be no clear sense in which the fact that $P(Q, q_i, S, t) = p$ is 'displayed' by what happens in S at t even if the enabling condition for the propensity, viz. that Q is measured on S at t, is realized. And indeed, to use frequentist terms, the display takes place over an ensemble of systems, similar to S at t, and on which Q is measured. In short the propensity must be seen as a *generalized* disposition, a disposition which is not displayed on each occasion that its enabling conditions are realized, but which is displayed by a statistical relation between its enabling conditions and its result. (For this view see Mellor (1971) pp. 68–70.)

Moreover, and here is the point to introducing this notion of a propensity as a generalized disposition, I shall take the state of S at t (as described by $f(S, t)$) to be the categorical basis for this propensity in just the way that an ordinary disposition has a categorical basis, i.e. in just the same way that we take the breaking of a piece of glass (when it is dropped) as a display of a disposition which the glass has, viz. the disposition to break under mild stress (its 'fragility'), and take the molecular structure of the glass to be a categorical basis for that disposition. In short I take it that the propensity for the measurement of Q in S at t to have result q_i plays a mediating role between the fact that $P(Q, q_i, S, t) = p$ and the state of S at t (as represented by $f(S, t)$), which is similar to the mediating role that the glass's fragility (its disposition to break) plays between the fact of the glass breaking on some occasion and its molecular state or structure. By assimilating the relation between the $P(Q, q_i, S, t)$ and $f(S, t)$ to an already familiar metaphysical relation (the relation of a disposition to its categorical basis) in this way, we explain the respective roles of $f(S, t)$ and $P(Q, q_i, S, t)$ within QT, i.e. we 'interpret' one aspect of the theory.

Note that the metaphysical scheme being advanced here as an interpretative scheme for QT is not taken to be independent of the theory being interpreted. On the contrary we have already modified the metaphysical scheme by introducing an extension to the notion of a disposition by allowing dispositions to include propensities. But neither is it the case that the metaphysics is simply parasitic on the theory, i.e. although the metaphysical schema of categorical basis and disposition has some flexibility it is also to a large extent historically determined. In short the process of interpretation of a theory like QT is to be seen as involving an adjustment between the categories of the understanding (in Kant's sense) and the categories of the theory. And as we have already seen this is a complex process, involving first of all a philosophical investigation and critique of the categories of the understanding themselves, e.g. arguments between a frequency and single-case interpretation of objective probabilities, but also involving an exploration of the theoretical categories, e.g. what are the bridge laws, and finally an adjustment of these two categorical levels to each other. (This final adjustment may in fact be two-way, although we have not yet here considered ways in which the theoretical categories may be adjusted to the metaphysical. Later however we shall consider clear cases of this, e.g. the hidden variables programme of QT.)

The identification of $f(S, t)$ as a description of the categorical basis for certain propensities which S at t has presupposes several facts about $f(S, t)$: first that S at t having a particular state-vector is a real property of S at t (not just a theoretical fiction), second that it is a property of the individual S at t, and third that this property has causal efficacy. Each of these three facts is controversial. Thus, as we have seen in Chapter 1, Bohr disputes the reality of the property of having a particular state-vector (taking an instrumentalist attitude to the state-vector formalism of QT); and Einstein, while accepting the reality behind the state-vector formalism, sees state-vectors as describing properties of ensembles of systems rather than individual systems, and thus would seem to see quantum effects as merely 'co-operative phenomena' emergent from the behaviour of collections of q-systems rather than manifest in the behaviour of an individual system; and of course the third fact is controversial if the first one is. In the following sections the arguments for each of these facts is presented in turn, thus affirming the proposal to see $f(S, t)$ as describing a categorical basis for the propensities which S at t has and which are displayed by the various $P(Q, q_i, S, t)$.

Finally in this section let me note that there is a more direct way of taking the connection between $P(Q, q_i, S, t)$ and $f(S, t)$ to be like the connection between a disposition and its categorical basis. Suppose we take (D) above as an interpretation of $P(Q, q_i, S, t) = p$. This then enables us to move straight from:

$$P(Q, q_i, S, t) = p$$

to the existence of a disposition, viz. the disposition implicit in the modal condition (D). It is this disposition which we now call 'a propensity', and it differs from ordinary dispositions like fragility only in so far as its display (what the disposition is a disposition for) is statistical. And this disposition can then be straightforwardly assumed to have a categorical basis, just as for any other disposition which is legitimated by a counterfactual conditional of the form 'were situation S to take place at t then result of type E would occur'. And it is *this* categorical basis which we can then take to be described by $f(S, t)$.

Taking the latter view has the advantage of conservatism, in that we do not need to generalize our notion of a disposition to include Mellor's propensities; but it has the disadvantage of lack of generality in that we need to restrict ourselves to (D) as an analysis of '$P(Q, q_i, S, t) = p$'. Further discussion of this issue takes us well beyond what we can, or need to, discuss here.[13] All I shall require, for later purposes, is the minimal claim that the probabilities of QT are displays of propensities for which the states of q-systems, as represented vectorially, are categorical bases, but I shall leave open whether those propensities are ordinary dispositions or generalized dispositions of the type Mellor proposes.

---

[13] See Mellor's discussion (1971) pp. 68 ff. for a prologue to this discussion.

5. BRIDGE LAWS

In this section we shall discuss how QT makes the connection between the 'observable' (or phenomenal) $P(Q, q_i, S, t)$ and the 'theoretical' $f(S, t)$. This discussion will reaffirm the suggestion made in the previous section that we treat $f(S, t)$ as a categorical basis underlying the various $P(Q, q_i, S, t)$.

With every pair (Q, q_i) (for Q non-degenerate) QT associates a unique (or nearly unique)[14] vector $f(Q, q_i)$—called 'the eigenvector of Q for eigenvalue q_i.' This $f(Q, q_i)$ is such that $f(S, t) = f(Q, q_i)$ iff $P(Q, q_i, S, t) = 1$, i.e. $f(Q, q_i)$ is the state of S at t which makes it certain for Q to be measured to have value q_i. QT then tells us that for any Q, q_i the probability of measuring Q to have value q_i in S at t is the square of the magnitude of $(f(Q, q_i), f(S, t))$, where $(f(Q, q_i), f(S, t))$ it will be remembered is a measure of the closeness of f_i to $f(S, t)$. More formally the bridge law is:

$$P(Q, q_i, S, t) = |(f(Q, q_i), f(S, t))|^2.$$

In other words the further away $f(S, t)$ is from a state in which it is certain that Q has value q_i the smaller is the probability that Q is measured to have value q_i in S at t. This law is called the Born statistical interpretation, '(BSI)' for short. The form of (BSI) clearly shows why it is only the relative orientation of pairs of vectors in $\mathcal{H}(S)$ which is of significance in QT. It is because only the relative orientation figures in (BSI) and hence affects the observational quantities $P(Q, q_i, S, t)$.

From the theoretical $f(S, t)$ and with the help of the $f(Q, q_i)$ which are supplied by QT we can always infer the observational $P(Q, q_i, S, t)$ via (BSI). Moreover it is easy to see that the converse inference holds too *if* there are enough of the $f(Q, q_i)$ for a given S, t; i.e. from $|(f(S, t), f(Q, q_i))|^2$, which is just $P(Q, q_i, S, t)$, by (BSI), we can tell to within a phase factor what the angle between $f(S, t)$ and $f(Q, q_i)$ is. But from the angles between $f(S, t)$ and enough fixed directions in the Hilbert space we can work back to what direction $f(S, t)$ is. In short given enough $f(Q, q_i)$ we can work back from the $P(Q, q_i, S, t)$ to $f(S, t)$. Thus at least under certain conditions (BSI) provides a two-way bridge between the theoretical $f(S, t)$ and the observable $P(Q, q_i, S, t)$.

---

[14] Actually, the $f(Q, q_i)$ are not quite unique, but all differ trivially from one another by a 'phase factor' in such a way that $|(f(Q, q_i), f(S, t))|^2$ is not affected by the difference.

This issue of whether, and under what conditions, (BSI) constitutes a two-way bridge law is clearly important for empiricists. If it is two-way then $f(S, t)$ itself becomes 'observable' (indirectly so at least) because it is deducible (via theory) from (directly) observable facts. On the other hand if the bridge law is one-way then the $f(S, t)$, like individual atoms, become inferential or theoretical entities, useful to bring forward in explanations, but not observable (even indirectly). And these epistemological classifications ('observable' and 'theoretical') are then taken by the empiricist to justify ontological claims (it is the observational entities which exist). We will not here be adopting this empiricist point of view on settling ontological issues. Nevertheless we note the issue of whether the bridge law is two-way is significant, because it will figure in connection with other issues later. In particular it will re-emerge as the issue of whether *any* given vector in $\mathcal{H}(S)$ is an eigenvector of some q-quantity (see Chapter 7).

What is of more immediate interest to us here, however, is that the bridge law connecting $P(Q, q_i, S, t)$ and $f(S, t)$, whether one- or two-way, provides support for the view advanced in the last section that $f(S, t)$ is a categorical basis for the propensities displayed by the $P(Q, q_i, S, t)$. This is because $f(S, t)$ together with the laws of QT entail what values the $P(Q, q_i, S, t)$ take, and this is exactly what we would expect for $f(S, t)$ to be a categorical basis for the propensities displayed by the $P(Q, q_i, S, t)$, i.e. we already here have some grounds for asserting the causal efficacy of state-vector assignments (although these grounds are by no means decisive).

6. THE STATE-VECTOR AS A SINGLE-SYSTEM PROPERTY

In this section I shall address not only the question of whether $f(S, t)$ represents a real property, but also the further question of what it is a property of, in particular whether it is a property of the *individual* system S at t. This question has been left open so far in part because of an already noted ambiguity in the interpretation of $P(Q, q_i, S, t)$. That is if we interpret $P(Q, q_i, S, t)$ in the first fashion indicated above, as a property of an ensemble of systems prepared similarly to (in the same state as) S at t, then it would seem natural to interpret $f(S, t)$ as representing a property of an ensemble of systems rather than of just S and t. On the other hand if we see $P(Q, q_i, S, t)$ as a single-case probability, and hence as the property of the individual S and t, then this inclines us to see $f(S, t)$ as representing a property of the individual S and t as well.

There is however another way of approaching this issue of the status of $f(S, t)$ independently of how we interpret $P(Q, q_i, S, t)$. What we will show below is that not only can we manipulate the state-vector of a particular system, but that the manipulation has effects which are discernible in the behaviour of the particular system, and are not just discernible as changes in statistics over an ensemble of similarly manipulated systems. These effects give reason for taking $f(S, t)$ as representative of a property of the individual S and t, rather than just as a property of an ensemble of systems and times prepared similarly to S at t. It also indicates that the property of having a particular state-vector is a real property and not just a theoretical fiction as Bohr regarded it. Here I am appealing to Hacking's criterion for reality: an entity is real if it can be kicked (manipulated) and kick back (have effects). (See Hacking (1983) pp. 272ff. for a defence of the latter criterion of reality against empiricist criteria which single out the explanatory power of entities, or their observational status, as grounds for taking them as real. Note however that since it is also the case that $f(S, t)$ is observational, with the qualification noted in section 5, *and* has high explanatory power in virtue of its role within QT, we see that its 'reality' is attested to by all of the above criteria.) In other words I am claiming here that it is because the property of having a particular state-vector can be kicked and kicks back at the level of the behaviour of S alone that we are justified in seeing it as a real property of the individual system S rather than a fictional property or a property of an ensemble of similar systems.

That the state-vector of S at t can be kicked, i.e. is manipulable, is clear enough. We simply change the state-vector for S at t by changing the boundary conditions on S at t: to put S at t in a square potential well, take it out of a square well, change the slope of the walls of the well, etc. all induce changes in $f(S, t)$. But how do we show that the state-vector of S at t kicks back? There is a trivial way in which $f(S, t)$ can be seen as having an effect on the behaviour of S at t. Consider a beam of electrons each in an eigenvector of Q for value q_1 at t; and suppose that the state-vector of one of them—say e— is switched to an eigenvector of Q for eigenvalue q_2 at time t', where $q_2 \neq q_1$. The measured values of Q over the original beam of electrons at t will be restricted to just q_1; in particular q_2 will never appear as a measured value. But the measured value of Q for e at t' is q_2. Hence the change of state-vector of e at t' has an observable effect (measured value $= q_2$), an effect which would not—indeed could not—have occurred had the state-vector of e at t' stayed as the eigenvector of Q for eigenvalue q_1.

But matters are not quite as straightforward as this last example suggests, because what we are considering there is a special case of measuring a physical quantity on a system which is in an eigenvector of the measured quantity, and hence the measured physical quantity can uncontroversially, even by Bohr's standards, be taken as having a particular value which the measurement simply reflects. As such the change in measured value of Q for e at t' can be seen simply as an effect of a change in the value possessed by Q for e at t'. This points to a difficulty for Hacking's criterion: it may be easy enough to establish the existence of some effect such as a change in measured value but not so easy to identify what is the cause of that effect from among a suite of theoretically related factors. Thus is it the state-vector of e at t' which kicks back by producing a measured value q_2 for Q, or is it the value possessed by Q which does the kicking?

I shall now consider a different experimental set-up in which these problems can be resolved. Reconsider the double-slit experiment, but now suppose that there is a preliminary collimation of the beam of electrons incident on the double-slit which puts the electrons arriving at the slits in a new state f which is not the momentum eigenstate f_p which we used in Chapter 1 to produce the interference effects. If the collimation is strict enough, to within an interval substantially less than the separation of the two slits, then, as we can observe, the interference effects vanish, i.e. a simple blurred image of two slits appears on the photographic plate.

This last fact may not at first sight seem to be a phenomenon to do with the behaviour of individual particles, but rather to do with their group behaviour. But consider more closely the beam of electrons for which each electron is in the state f when arriving at the slits. Some of these electrons eventually land on the plate at places where they could not have landed had they been in the state f_p initially. In particular this will be so for the electron—call it 'e'—which lands on the plate at a null point of the interference pattern which we discussed in Chapter 1. By landing on a null point, e is precisely exhibiting an effect of its being in the state f rather than f_p; i.e. it goes where it could not go had it been in the state f_p. (Note that this argument trades on an assumption which we gave up earlier on, viz. that we can tell whether an electron lands precisely at a particular point on the plate. But this whole argument can be restated for discrete valued physical quantities, without making this assumption.)[15]

The phenomenon we are confronted with here—of *e* being caused to land at a null point of the interference pattern by a change of state-vector—is exactly the sort of singular causal relation which, we claimed, justifies taking a state-vector for *S* at *t* as representing not just a real property, but a real property of *S* at *t*. Moreover we see that the ambiguity which was present in the phenomenon we considered earlier is not present here. That is there is no question of simply attributing *e*'s ending up at a null point of the interference pattern to a change in *e*'s trajectory, because it is precisely in the double-slit experiment that the assumption of a particular trajectory strikes difficulties, as we saw in Chapter 1. Thus it is the state-vector of the electron *e* which stands out as the cause of the change in *e*'s point of termination on the plate. And so, by the above criterion, we conclude not only that the state-vector represents a real property, but a real property of the particular system *e*.

[15] Suppose the initial electrons are all in a state orthogonal to the eigenvector of Q for value q_1. Then the probability of measuring Q to have value q_1 is zero. Now switch the electrons to a state orthogonal to the eigenvector of Q for eigenvalue q_2, and let e be that electron which then happens to register the value q_1 for a Q-measurement. This can all be done without letting the relevant state be an eigenvector of Q.

3

The Pilot-field

The pilot-field interpretation; probability fields; collapse of the
wave-packet; the Stern–Gerlach experiment; quantum systems
as particles; non-locality.

1. THE PILOT-FIELD INTERPRETATION

In the previous chapter we described the core features of the
theoretical role which the state-vector has in QT (sections 4 and 5),
and said what a state-vector is (sections 1–3). Moreover we located
the property of having a state-vector as a real property of an
individual system at a time (section 6). And we went some of the way
to interpreting the state-vector by locating it within a suite of
metaphysical models in terms of which explanations are couched
(section 5). In particular we suggested that $f(S,t)$ describes a categorical basis for the propensities displayed by the $P(Q, q_i, S, t)$. In this
chapter I shall complete the task of interpreting the state-vector. In
particular I shall argue that $f(S, t)$ should be taken as a field-function,
describing a probability field which is associated with S at t.

Aspects of this suggestion are by no means novel. Indeed it is in
general terms a return to an interpretation, which preceded Danish
orthodoxy, in which q-systems are particles associated with fields.
The earliest form of this interpretation was put forward by de Broglie
in 1923 who referred to the field associated with a q-system as a
'fictitious associated wave'.[1] It seems to have been independently
suggested by Einstein,[2] and was taken up by Schrödinger in 1925, to
be formalized in his Wave Mechanics paper of 1926 (although
Schrödinger never made the transition to a *probability* field, being

---

[1] De Broglie (1923).   [2] See the discussion in Pais (1982), ch. 24, sect. C.

content to see the electron's field as a distribution of electronic charge).[3] Born took this idea further by introducing the idea of a probability field in his 1926 papers,[4] taking it as explicating Einstein's unpublished notion of a pilot-field.[5] One can date the inception of Danish orthodoxy from Bohr's Como lecture of 1927.[6] The pilot-field interpretation effectively vanished from that date, except for a revival by Renninger in 1953 which drew positive responses from both Einstein and Born.[7]

If what I am suggesting here is a return to a superseded interpretation then it is important to understand why it was superseded in the first place, in order to ensure that the older objections to it can be circumvented. But here we strike a surprising fact. The historical reasons for its demise are hard to trace. In part it seems to have been because the whole idea of a field which was displayed statistically conflicted with strong metaphysical convictions that nature was deterministic—Einstein's 'God does not play dice'.[8] In part it seems to have been because the pilot-field turned out to be non-classical in its form of representation, as indicated by Einstein's disparaging description of this as 'Waves in $3n$ dimensional space whose velocity is regulated by potential energy, e.g. rubber bands'.[9] But neither of these reasons is compelling today. Indeed, somewhat ironically, it is Bohr's rival interpretation of QT which has rendered the first of these reasons void (by teaching us to embrace indeterminism). Moreover, the forms of representation for fields in other areas of physics have undergone radical changes from their classical Newtonian–Maxwellian paradigms, e.g. the tensor fields of general relativity put forward by Einstein in 1912 and published in

---

[3] See Schrödinger (1926a) and (1926b). For a history of Schrödinger's views see McKinnon (1980) and Wessels (1980).

[4] See Born (1926a) and (1926b).   [5] See Pais (1982) p. 441.

[6] Reprinted in Bohr (1928)—see the discussion in Jammer (1974) p. 86ff.

[7] Einstein wrote to Renninger: 'One has to ascribe real existence both to the wave field and the more or less localized quantum', and Born wrote: 'Your view is precisely the one I have espoused . . . both particles and waves have some sort of reality'—see Jammer (1974) pp. 494 and 495.

[8] See Pais (1982) pp. 412 and 443. Note here the taken for granted opposition between determinism and fundamental laws being probabilistic which I shall criticize later.

[9] Pais (1982) p. 443. n here is the number of systems so that it is multiple systems (with $n \geq 2$) for which the fields cannot be represented by a function of position in three-dimensional physical space. Also see Wessels (1980) p. 68.

1916,[10] and so the second reason given above no longer seems cogent either (nor should it have been all that cogent at the time).

A reason for the decline of the pilot-field model can also be found in the traditional interpretation of probabilities as either subjective expressions of our ignorance or as relative frequencies[11], and hence, on either view, not the objective properties of *single* systems, as they would surely have to be if a probability field were to be an objective entity associated with a single *q*-system. But this reason too is no longer cogent given the third way of interpreting probabilities as displays of objective single-case propensities discussed in Chapter 2. In short it seems that none of the historical reasons for rejecting the pilot-field interpretation still applies. And this leaves open the possibility of a return to this interpretation. Indeed more than that, I shall argue, such a return is now an appropriate move in order to set right some of the metaphysical excesses of Bohr's metaphysics.

The usual sorts of fields described in classical physics, e.g. in classical electromagnetism and Newtonian gravitational theory, are characterized in terms of some physical quantity, such as force, energy, or some potential, taking a value at each point in an extended spatial domain. Moreover the fact of this physical quantity taking a particular value at a particular point in that domain is taken to cause a certain dispositional effect at that point, e.g. an acceleration of a certain magnitude and direction on certain 'test particles' were they to be introduced into the field at that point. The field can thus be seen as the seat—indeed the categorical basis—of these various dispositions. Moreover there need be no material object or substance in the spatial domain of the field. For example the electromagnetic field may exist *in vacuo*. In short the field is conceived of as a not necessarily substantival categorical basis for a spatial distribution of dispositions, and is characterized in terms of a field-function which details the spatial distribution of values for the relevant field quantity (be it force, energy, or whatever).[12]

I shall here suggest that associated with any *q*-system *S* at any time *t* is a field, described by $f(S, t)$, which is a categorical basis for certain spatially distributed propensities. The propensities in question are propensities for *S* to be measured at various locations at *t*, and are

[10] See Pais (1982), ch. 12, sect. 6.

[11] For a discussion of this traditional dichotomy see van Frassen (1980), ch. 6, sect. 6.

[12] For a general discussion of the nature of fields, see Hesse (1961).

displayed by there being a certain probability of measuring S to be at x at t for various x. The value of this probability can be deduced from the state-vector of S at t at x via the Born statistical interpretation. In this way the state-vector plays the role of the field-function. The difference from classical fields is twofold however. First the method of description is not in terms of the spatial distribution of values for some physical quantity, but in terms of a state-vector, and second the propensities for which the field provides the categorical basis are displayed statistically.

Note that the field $f(S, t)$ cannot in general be completely characterized as a basis for the propensities for position measurements of S to yield a positive result at various space points. In particular we may have $f(S, t) \neq f(S, t')$ and yet the propensities for S to be found at x at t for varying x are the same as they are for S at t'. The difference between $f(S, t)$ and $f(S, t')$ in this case will be manifested by certain of the $P(Q, q_i, S, t)$ differing in value from the corresponding probabilities for S at t', but the probabilities here will not be the probabilities for a position measurement to reveal certain values.[13] Thus completely to characterize $f(S, t)$ we must take it as a basis for propensities other than propensities for position measurements to have certain results.

Note also that in the case of a single q-system the state-vector can be replaced by a function of position—the 'state-function'—which describes the field in an equivalent way to the state-vector, as we shall see in the next chapter. And so at least for single q-systems the classical idea of a field-function which is a function of position can be retained. But for systems which consist of multiple q-systems this replacement is no longer possible: the state-function for a multiple q-system is a function of several different position variables.

Both of the preceding notes indicate that if we are to see the state-function as a field-function then it cannot simply be as a function of position. In QT therefore it seems that the notion of a field must be subjected to two novel generalizations if it is to be of any use. First of all we must allow talk of a field as a categorical basis for propensities which are displayed statistically, and second we must allow a more

---

[13] This is obvious if we switch to a representation of $f(S, t)$ as a function of position. Then the probability of finding S at t within dx of x is proportional to the magnitude of the value of $f(S, t)$ at x squared times dx. But clearly the squares of the magnitudes of the values of $f(S, t)$ and $f(S, t')$ at x may be identical and yet $f(S, t) \neq f(S, t')$.

abstract description than in terms of a function of position, in particular we must allow a vector representation or equivalently a function of several position variables in the case of multi-q-systems. Indeed later I shall suggest that the field be represented even more abstractly by an operator on a Hilbert space, viz. the density operator.

What I have suggested so far is that there is a field $f(S, t)$—what Born calls 'a probability field'—associated with the q-system S at time t. Although Born's term 'probability field' is a useful term here, it is important to note one point at which this term is quite misleading. Consider a Newtonian gravitational field. To call this 'an acceleration field' would be quite misleading. The field is not a field of accelerations. It is a field of forces, which are categorical bases for the acceleration of test particles were any to be inserted in the field. The accelerations display the presence of the field but do not constitute it. In the same way it is misleading to say that the field $f(S, t)$ is a probability field. It is not a field of probabilities or even of propensities (any more than a force field is a field of tendencies). Rather it is a categorical basis for propensities, the presence of which is displayed by the probabilities of certain hypothetical experiments to have certain results. For convenience however I shall retain the term 'probability field' in what follows.

There are several ways in which the suggestion that $f(S, t)$ describes a field associated with S at t may be interpreted. One of these ways we have already mentioned: to see the field as a pilot-field, which steers the course of the particle S but is nevertheless distinct from S itself. It is characteristic of the way this pilot-field view was constituted historically that S is seen as having a unique determinate position at any one time, viz. one of the positions at which the pilot-field has non-zero amplitude. Thus there is a conceptual distinction made between the location(s) of the field associated with S at t and the location of S itself at t. Moreover this is a 'hidden variables' view in that the position of S at t is taken to be always determinate, albeit simply one of many positions at which the pilot-field has non-zero amplitude.

A second way is to see S at t as a metaphysical hybrid—a permanent cross between a particle and a field—with both a field aspect (as described by $f(S, t)$) and a particle aspect. On this view the 'field' associated with S is fictional—an idealization, which cannot be conceptually separated from S (in the same way that 'justice' cannot be separated from just men). But as we shall see this view has difficulties handling the Stern–Gerlach experiment; in particular it does not seem able to explain how the probability field associated with a silver ion is split into two by a magnetic field (a split demonstrated experimentally by the two halves interfering with each other), while nevertheless the ion remains whole, i.e. is deflected either up or down by the field and not split into two 'ghost ions'.

A third way in which the relation between S at t and its associated field can be seen is to take the relation as one of identity, i.e. the field described by $f(S, t)$ *is* S at t. Under this interpretation the 'position measurements of S' are reinterpreted as localized particle detection experiments for which the detectors are triggered by the field S with probabilities which depend on where the detectors are located in the field. A positive result for such a detection experiment does not however mean that a particle is actually located at the point of detection; it simply means that the field S has managed to trigger off the detector at the point of detection. Indeed the field S is regarded as *inter alia* a categorical basis for tendencies (propensities) to trigger off just such localized particle detectors. This suggestion does have an advantage of ontological simplicity. As we shall see however it does not do justice to the full range of quantum phenomena. In particular it leaves mysterious the particle-like aspect of q-systems—why they are observed at only one place at any one time.

In the rest of this chapter I shall argue for a slightly modified version of the first of the above options, viz. Born's pilot-field model for q-systems. The modification I suggest will be to allow the particle which is being piloted to have an indeterminate location and not simply sit at a determinate location inside the pilot-field. Moreover, as already indicated, I shall differ from the very early versions (de Broglie's) of the pilot-field model in taking the pilot-field as real (*qua* not fictional)—indeed as a probability field (thus going beyond Schrödinger too)—but with the probabilities interpreted as displays of propensities. In order to argue for this modified pilot-field interpretation, I shall point out its advantages not only over the other two 'field interpretations' of q-systems (q-systems as particle–field hybrids and the q-systems as just fields) given in the previous paragraphs, but also over the orthodox view and Born's original pilot-field interpretation. (Note that the modification of Born's original views which I shall suggest here—to see the particle as not precisely localized inside its pilot-field—was in fact taken for granted by Einstein in his letter to Renninger in 1953—see footnote 7 above. Einstein however would never have accepted the notion of a probability field, which I take from Born. Thus what I am putting forward here is a hybrid of Born's and Einstein's views.)

First, I shall contrast the modified pilot-field interpretation with the orthodox interpretation. On the orthodox interpretation the characteristics deployed by a q-system at some time—wave characteristics or particle characteristics—depend on the experimental conditions at that time. By contrast, on the pilot-field model, a q-system may be said to have *both* of these families of characteristics under *all* conditions, i.e. the q-system is a particle always, but is also always associated with a field. Thus on the latter view we return to a unified, *qua* temporally invariant, metaphysical picture for q-systems, albeit one which is a composite, indeed a simple juxtaposition, of earlier pictures. In particular we no longer need to assume that as experimental conditions change, so do the metaphysical pictures in terms of which q-systems are described. (Although note here that Born, in the context of putting forward a pilot-field model for q-systems, argues that Bohr was misunderstood on this last point.)[14]

There is a further important point on which the pilot-field model for q-systems differs from the orthodox view. On the orthodox view the particle picture proposes a system with a unique determinate position, whereas the wave picture is of a system with an indeterminate position. Indeed it is the indeterminate position which is implicitly taken to explain the wavelike behaviour of q-systems, i.e. the q-system is taken to produce interference effects in the way that a spatially extended wave does, just because its position is indeterminate. More particularly it is the region within which the q-system's position is indeterminate which is taken as the region over which the q-system is taken to be extended. By contrast, on the pilot-field model it is the probability field associated with S at t, not its indeterminate position, which is responsible for its wavelike behaviour.

The view advanced by the pilot-field model on this last issue is borne out by two considerations. First, if we look at the double-slit experiment then we see that the particular form taken by the interference pattern is explained not by the indeterminacy in position, but by the form of the state-function.[15] Indeed (as we saw in Chapter 1) the indeterminacy in position only serves to explain the flaw in one particular argument which purports to demonstrate that the double-

---

[14] Born (1949) pp. 103 and 105.
[15] See Feynman (1965), ch. 1 for example.

slit pattern must be a linear combination of the two single-slit patterns; but that is quite different from explaining the characteristic interference pattern which is produced in the two-slits-open part of the experiment. To explain the latter we need to appeal to the wave form of the state-function, and that, in our terms, means the wave form of the probability field. Thus it is the probability field, not the indeterminacy of position, which explains the interference effects characteristic of the wave picture.

This same conclusion is even more strongly indicated by the results of the Stern–Gerlach experiment which, as we shall see shortly, show interference effects for a stream of silver ions each member of which has a quite determinate position. Thus whatever is responsible for the wavelike behaviour of these silver ions, it cannot be their indeterminacy of position.

In short we see that there are two points at which Bohr's views differ from the pilot-field model for q-systems. First, for Bohr a q-system is either totally a wave or totally a particle, depending on the experimental conditions (or, in less tendentious terms, we might say that at times a q-system either totally satisfies a wave picture *or* totally satisfies a particle picture).[16] On the pilot-field model however, a q-system is always already a bricolage of both wave, *qua* field, and particle. Second, when Bohr refers to 'wave' and 'particle' he means classical waves and classical particles. But in the pilot-field model for q-systems, the q-system is not so much a (diachronic) combination of a classical wave and a classical particle, as a (synchronic) combination between what I have called 'a probability field' and a particle. Moreover, on the modified pilot-field model which I shall be suggesting, the particle in question may have an indeterminate position. Therefore a q-system S, at any time t, is seen as a combination of two non-classical entities: a probability field and a (possibly) fuzzy particle (with indeterminate location). The neologism 'fieldicle' is perhaps appropriate here (in parody of the familiar term 'wavicle'), as long as it is kept firmly in mind that (a) the field is a probability field not a classical wave, and (b) the field and particle natures are combined synchronically—juxtaposed—rather than being alternated diachronically as they are on the orthodox interpretation.

---

[16] Feyerabend (1967) argues for the former reading of Bohr's views.

It is also important to appreciate how the modified pilot-field model I shall be suggesting here differs from the original pilot-field model put forward by Born, in which the step to a probability field was first made. In addition to the conceptual differences (with regard to the interpretation of probabilities), there is the difference that the Born pilot-field model basically took q-systems to be classical particles in respect of having determinate positions (indeed determinate values for any q-quantities), their only difference from classical particles arising dynamically because they happen to be steered by probability fields. This view therefore confronts all the difficulties which hidden variables theories have subsequently been found to confront.[17] And also of course it has difficulties in avoiding the conclusion of the argument in Chapter 1 that the two-slit pattern is a linear combination of the one-slit patterns. The modified pilot-field model allows that the position of q-systems may be indeterminate, and hence does not confront these difficulties.

The Born pilot-field model also differs from the modification I am suggesting on the related point that it (the Born model) explicates the apparent 'indeterminacy' in position of q-systems as mere *uncertainty* in position. In particular it identifies the range of indeterminacy for S at t with the spatial domain over which the field-function associated with S at t is non-negligible.[18] But we shall demonstrate below that this identification fails for Stern–Gerlach silver ions which have dispersed probability fields, despite having unique determinate (determinate but uncertain) positions on emerging from the magnetic field. (This identification fails, as we shall see, for any q-systems like an electron for which the Hilbert space is a direct product of two subspaces, on one of which the position X of the q-system is defined, on the other of which some other q-quantity Q, such as spin, is defined. It is this decomposition of the Hilbert space which allows the production by measurement of Q of an entangled state for that system, for which X has determinate value, even though it is in a superposition of X eigenvectors. Note that the identification also fails in cases for which systems are in mixed states but are nevertheless parts of combined systems which are in pure states—as in the case of

---

[17] See Chapter 8 on Stapp and Eberhard.
[18] This identification is apparent in those 'derivations' of the indeterminacy principle in which the indeterminacy of position for some q-system is identified with an indicator of the spread of the associated wave-packet. See Heisenberg (1930) p. 14 for example.

an electron which belongs to an Einstein–Podolski–Rosen electron pair. The formal details of this will be clarified in Chapter 4.)

The argument so far: I have indicated disadvantages which both the orthodox and Born pilot-field interpretations have *vis-à-vis* the modified pilot-field interpretation. The main disadvantage centred on the failure of the former interpretations to account for interference effects shown by Stern–Gerlach silver ions which have a determinate position. (An account of the Stern–Gerlach experiment will be given in the next section.) This still leaves two other rival interpretations to criticize: the *q*-system as identical with its associated field, or as a particle–field hybrid. I shall criticize these in the final section of this chapter, but in the next section I shall address another essential aspect of the argument for the modified pilot-field interpretation, viz. a defence of it and related field interpretations of *q*-systems against a criticism based on the collapse of the wave-packet.

2. COLLAPSE OF THE WAVE-PACKET AND THE STERN–GERLACH EXPERIMENT

At first sight the view of $f(S, t)$ as describing a probability field associated with S at t is subject to precisely the same difficulties as the wave model, namely the difficulties from the collapse of the wave-packet phenomenon. Consider S at t when it has non-zero probability of being detected at the location x and also has non-zero probability of being detected at location x' widely separated from x. Indeed we can choose S at t such that x and x' are as widely separated as we please. This means that the field which we associate with S can be chosen so that at t it extends over as large a distance as we please. But suppose a position measurement of S is started at t so that a positive detection of S at some location or other has taken place by the time t' shortly after t. This measurement of position is traditionally taken to be accompanied by a 'collapse of the wave-packet', i.e. all of the field collapses down to the point (or small region) at which S is detected at t'. But this in turn means that the edge of the field travels an arbitrarily large distance in the short time $(t' - t)$ needed for the measurement to be completed. In other words the field velocity is arbitrarily large. But the velocities of physical systems are generally taken to be bounded above by the velocity of light, and hence we have a contradiction.

There are several reasons why this argument is not compelling. Firstly, the principle that entities do not have superluminal velocities is only justified (from within special relativity) for entities which have a real rest mass. But the collapse of the wave-packet phenomenon does not involve a mass transfer since the probability field does not consist of a mass (or indeed anything substantial) existing at every point in the field. And so a superluminal collapse of the probability field is not after all ruled out by the above principle.[19]

My second criticism of the argument concentrates attention on its premiss that there is a collapse of the wave-packet on measurement. Since the issue of the collapse of the wave-packet is an important issue in its own right, independently of the particular context in which it is being considered here (of how to interpret $f(S, t)$), I shall spend some time now in looking at these criticisms. In particular I shall criticize three arguments which may be mounted in favour of the hypothesis of the collapse of the wave-packet, and then demonstrate (via a consideration of the Stern–Gerlach experiment) the experimental inadequacy of this hypothesis as well as its inconsistency with the laws of QT. These considerations together make it clear that none of the field interpretations for $f(S, t)$ has anything to fear from the argument at the beginning of this section; nothing to fear because one of the argument's premisses, viz. that there is a collapse of the wave-packet, is unacceptable. These arguments also show the implausibility of any interpretation of QT which incorporates a collapse of the wave-packet as part of the measurement process; and thereby pave the way to the theory of measurement which I shall be advancing later.

The collapse of the wave-packet can be, and often is,[20] argued for by committing what Popper calls 'the great quantum muddle'.[21] The great quantum muddle is to suppose, just on the basis of observing S at (or near) x at t', that the probability of S being measured at (or near) x' at t' is 0 for any $x' \neq x$, and is 1 for $x = x'$. If we make this muddled inference then it does indeed follow straightforwardly from (BSI) that the state-function for S at t' is zero everywhere except at (or near) x; and the collapse of the wave-packet has been established. But, as

---

[19] Cf. the quotation from Born in Jammer (1974) p. 495: 'the waves are not carriers of energy or momentum'.

[20] See for example Heisenberg (1930) p. 36: 'As our knowledge of the system does change discontinuously at each observation its mathematical representation must also change discontinuously.' But Cartwright, as we shall see later, argues for the collapse of the wave-packet along different lines.

[21] See Popper (1983c) pp. 64 and 76ff.

Popper points out (1983c), we can question the muddled inference principle which seems to be at work here:

(Inf)   Experiment E has the result R at t'

∴ the probability of E having result R at t' is 1.

This inference seems to be invalid because of the following apparent counter-example. Let E be a particular toss of a die which stops rolling at t', and let R be the result of coming up six. Then $P(E$ had R at $t') = 1/6$ if the die and toss are fair (by definition of 'fair'). And this is so whether or not E in fact has the result R at t'. This same point (the invalidity of the above inference, and hence the failure of one way of justifying the collapse of the wave-packet) is made by Cartwright (1947) p. 233, and she locates the same point in the works of Margenau, Groenewold, Ballentine, and van Frassen.

There may however be some who disagree with Popper's (and Cartwright's) line of argument here, by arguing that if $P(E$ has R at $t')$ is not a frequency probability but rather some sort of objective single-case chance (and this is certainly how I have construed the relevant quantum theoretic probabilities) then $P(E$ has R at $t')$ is 1 or 0, depending on whether E has result R or not at t'. They base this conclusion on a claim of universal determinism—that all processes, including E, are deterministic—together with the principle that a process being deterministic entails that the objective single-case chances of its actual outcome is 1 and of any other possible outcome is 0.[22] Thus it is only the frequency probability of coming up six for tosses which are as far as we can tell indistinguishable from the particular toss in question which takes the value 1/6. Moreover they claim that it is our confusion of the latter frequency probability with the former objective single-case probability which makes us erroneously assign the value 1/6 to $P(E$ has R at $t')$.

I reject the latter view however; and follow instead Mellor's later view that determinism does not preclude the existence of a non-trivial

---

[22] Mellor endorsed the latter principle in his (1971). Lewis also seems to endorse something like this in (1980) p. 226, where he takes the probabilistic nature of coin-tossing to depend on the intrusion of external factors (such as air resistance) into the process, i.e. on the failure of the process to be *locally* deterministic. Ironically from the point of view of the above argument, he turns the latter principle around, using it to argue from the probabilistic nature of coin-tossing to its being *locally* indeterministic (although this is of course consistent with *global* determinism).

single-case objective chance.[23] Note however that even if one does not follow Mellor's later view the above muddled inference (Inf) can be seen as invalid by questioning the implicit premiss of determinism. That is one need only substitute for E some uncontroversially indeterministic process which displays a non-trivial statistical distribution over its possible outcomes, e.g. the decay of a radium atom within a certain period of time, to see that (Inf) is invalid. And such a substitution is surely exactly to the point here, since the process we are considering (measurement in QT) is an indeterministic process *par excellence*.

There is however another way to argue for the collapse of the wave-packet purely on classical probabilistic lines, and which does not involve the above muddled inference. It uses a time-dependent objective single-case probability, unlike the previous argument which used an objective single-case probability $P(E$ has R at $t')$ which is only *apparently* time-dependent. ($P(E$ has R at $t')$ is only apparently time-dependent because the specification 'at t'' in the argument of the probability expression is redundant, i.e. it merely serves to identify the terminal point of E, an identification which is already implicit in the definition of E.) Lewis provides an example of just such a time-dependent single-case objective chance:

We ordinarily think of chance as time-dependent . . . Suppose you enter a labyrinth at 11.00 a.m., planning to choose your turn whenever you come to a branch point by tossing a coin. When you enter at 11.00, you may have a 42% chance of reaching the center by noon. But in the first half-hour you may stray into a region from which it is hard to reach the center, so that by 11.30 your chance of reaching the center has now fallen to 26%. But then you turn lucky; by 11.45 you are not far from the center and your chance of reaching it by noon is 78%. At 11.49 you reach the center; then and forever more your chance of reaching it by noon is 100%.[24]

(Lewis goes on to say that this paradigm of time-dependent single-case objective chance is the model for all objective single-case objective chances. In particular what appears to be a time-independent chance is merely a time-dependent chance considered in a context which implicitly fixes the time: 'We might well say (before,

---

[23] See Mellor (1981). In support of Mellor's view it is interesting to note Poincaré's famous explanation of the existence of a non-trivial probability distribution (for outcomes of roulette) depends on assuming determinism not indeterminism—see Poincaré (1952).

[24] Lewis (1980) p. 232.

after or during your exploration) that your chance of reaching the center is 42%. The understood time of reference is the time when your exploration begins.' We will not discuss this question of the priority of time-dependent probabilities here).

Moreover from Lewis's Reformulated Principal Principle[25] it is easy to see that the time-dependent probability, relative to time t, that E has result R is given by:

$$P_t(E \text{ has } R) = P_t(E \text{ has } R | H_t),$$

where 'H_t' stands for the history of the world up to t. (For convenience we are considering probabilities at the actual world.) Thus we see that the time-dependence of the probability at time t of result R for process E arises in part because the set of factors up to time t which influence the result of E changes as t changes. (Note that the appearance which $P_t(E \text{ has } R)$ has of being subjective only arises because we estimate it via the relative frequency with which E has R over what are, as far as we know, relevant re-runs of H_t.)

Now if we accept all of this then, from the Reformulated Principal Principle, it clearly follows that for any t at or after the end of E, $P_t(E \text{ has } R) = 1$ or 0, depending on whether E has outcome R or not. (This is because H_t must include whether or not E has R for any t at or after the end of E, and because $P_t(E \text{ has } R | H_t) = 1$ or 0 respectively if H_t includes the fact that E has R or the fact that E does not have R respectively.) Thus we have a collapse of the probabilities at time t', i.e. $P_t(E \text{ has } R) < 1$ for $t < t'$ but $P_t(E \text{ has } R) = 1$ or 0 for all $t \geq t'$. And applying this result to the toss of the die we see that indeed the probability, relative to time t', of its coming up six is 1 or 0, depending on whether it comes up six or not. (This does not of course contradict the earlier claim that the time-*independent* probability of it turning up six at t' is 1/6.)

And applying this result to the measurement process which ends at t' with S being measured at x we get:

$P_{t'}(S \text{ is measured at } x) = 1$ but
$P_t(S \text{ is measured at } x) < 1$ for $t < t'$.

Therefore if we interpret the phrase 'the probability of measuring S at x at t'', as it occurs in QT in (BSI), as '$P_{t'}(S \text{ is measured at } x)$' then we

---

do indeed get our collapse of the wave-packet at t'. Moreover we do so without invoking the muddled inference principle (Inf).

But there is still a muddle here, albeit a different one. The plate only measures the position of S at one time, viz. at that time when S starts to interact with the plate to an extent that leads irreversibly to a spot appearing on the plate. Suppose that this measurement interaction starts at t to end at t'. If we are implicitly referring to that particular measurement (from t to t') then indeed

$P_{t'}(S$ is measured at $x) = 1$.

More explicitly, and denoting that particular measurement by M, we get:

$P_{t'}(S$ is measured by M at $x) = 1$.

But this only implies a collapse of the wave-packet if the latter probability is the probability referred to in QT as 'the probability of measuring S at t' at x'. Clearly however it is not. The quantum theoretic probability of measuring S at t' at x is the probability of measuring S at x for a quite different measurement—call it M'—viz. one starting at t' (not at t) and finishing at some even later time t''. (The distinction between M and M' is difficult to make within the usual rhetoric of QT, because measurements tend to be treated as instantaneous so that the distinction between the temporal locations of M and M' is lost.) Thus formally we have:

$P_{QT}($measuring S at x at $t') = P_{t'}(S$ is measured by M' at $x)$.

And that last probability need not be 1, at least not purely on the basis of probabilistic arguments. Indeed it is 1 iff $f(S, t)$ is an eigenvector of position for eigenvalue(s) which are equal to (or near to) x.[26] And that in turn is so iff there is indeed a collapse of the wave-packet at t', which is of course exactly the point at issue. So it seems that after all, purely on the basis of probabilistic arguments, no argument for a collapse of the wave-packet can be mounted. Rather it is the collapse of the wave-packet which, if it occurs, provides reason for a collapse of the probabilities (if it occurs). And this is so even if we grant Lewis his time-dependent objective single-case chances.

There is a third argument which has been mounted for a collapse of the wave-packet for cases of conservative measurements. (Conservative measurements are those for which the value of the measured

---

[26] Or a mixture of such eigenvectors, see Chapter 4.

quantity is preserved by the measurement interaction.) An example of such a measurement is the Stern–Gerlach experiment, in which we magnetically split a collimated beam of silver ions.[27] A beam of silver ions pass in the x-direction between the poles of a magnet with north–south poles oriented in the y-direction. Each silver ion is deflected a macroscopically significant distance up or down in the z-direction depending on whether its y-component of spin (which is conserved by the interaction) is $1/2$ with certainty or $-1/2$ with certainty. Thus the interaction between each silver ion and the field causes a correlation between the silver ion's position (its z-displacement) and its spin (the y-component), in such a way that the events of having different y-component spin values with certainty leave macroscopically distinct traces in the form of differing z-displacements for the silver ions. It is because of this correlation that the Stern–Gerlach experiment is said to constitute a measurement.

Now consider those silver ions which have been displaced in the positive z-direction at some time t' after the interaction has stopped. Because the interaction is conservative it is tempting to suggest that at t' each of these silver ions is in a pure state which is an eigenvector of the y-component spin for eigenvalue $1/2$, i.e. in the state $f_{y, 1/2}$ for which y-spin has value $1/2$ with certainty. Indeed the usual rhetoric in terms of which we describe the Stern–Gerlach experiment, viz. as a 'state preparation', as a means of preparing some silver ions (those displaced in the positive z-direction) to have value $1/2$ for their y-component spin, affirms this suggestion. But this suggestion is just the collapse of the wave-packet hypothesis, that measurement collapses the state of a q-system into an eigenvector of what is measured (viz. into an eigenvector for an eigenvalue equal to the measured value). Thus the practice of physicists (how they see the Stern–Gerlach experiment) provides an argument for the collapse of the wave-packet. Cartwright (1983) p. 174 seems to take this argument as decisive: 'If we are to get our beam at the end of the accelerator, Schrödinger evolution must give out. Reduction of the wave-packet must occur . . . This kind of situation occurs all the time. In our laboratory we prepare thousands of different states . . . every day.'

Moreover subsequent measurements of those silver ions seem to bear this suggestion out. In particular a y-component spin measurement on any of them just after t' reveals the value $1/2$. And in addition

---

[27] See the discussion by Wigner (1967) pp. 160 ff. for example.

there are strong theoretical reasons for endorsing this suggestion, viz. that it resolves the Schrödinger cat paradox. (Cartwright makes what is essentially this same point in her (1983) p. 171, where she points out that the reduction of the wave-packet is needed to avoid 'the picture of objects jumping about' which she takes as 'crazy'. We shall discuss this last point in Chapter 5.) In short we see that there are good grounds for asserting the collapse of the wave-packet not only from the practice of physicists, but also from experimental results and theory.

This argument for the collapse of the wave-packet can however be undermined by attending to the results of measuring a physical quantity on the Stern–Gerlach silver ions which is neither compatible with the y-component spin nor the z-displacement, e.g. by measuring $p_z \times s_z$ the product of the z-component momentum and z-component spin. The results of this measurement show that the state of the silver ions at t' is 'entangled' in the following sense. (And here I introduce some formalism which may only become clear after the next chapter.) If the initial state (before entry into the magnetic field) of the electrons is

$$(c_1 f_{y, 1/2} + c_2 f_{y, -1/2}) \times f_{z, 0},$$

where $f_{z, 0}$ is an eigenstate of the z-displacement for $z = 0$, then at t' the silver ions are in the state

(i) $c_1 f_{y, 1/2} \times f_{z, +} + c_2 f_{y, -1/2} \times f_{z, -}$,

where $f_{z, +}$ is an eigenstate of z-displacement for large positive displacement and $f_{z, -}$ is an eigenstate of z-displacement for large negative displacement. (In particular the silver ions at t' are not in the 'reduced state' defined by the density operator:

(ii) $|c_1|^2 \hat{P}(f_{y, 1/2} \times f_{z, +}) + |c_2|^2 \hat{P}(f_{y, -1/2} \times f_{z, -})$,

which they would have to be were there probability $|c_1|^2$ that they are in state $f_{y, 1/2} \times f_{z, +}$ and probability $|c_2|^2$ that they are in state $f_{y, -1/2} \times f_{z, -}$.) And an ensemble of silver ions in this state can indeed be experimentally differentiated, by the results of measurement of $p_z \times s_z$, say, from an ensemble of silver ions some of which are in the state $f_{y, 1/2} \times f_{z, +}$ others in $f_{y, -1/2} \times f_{z, -}$.

Therefore we can experimentally verify that there is no collapse of the wave-packet; indeed the previous argument indicates that to suppose a collapse of the wave-packet is actually inconsistent with the laws of QT. Note that the inconsistency is derived here because it is accepted that the laws of QT apply to the interaction between the magnetic field and the silver ions right up to t'. There is in short no interval within the measurement interaction when the ordinary laws of QT are not taken to apply. By contrast, for other measurement processes, for which the nature of the measurement interaction is not known, our ignorance leaves room to hypothesize a collapse of the wave-packet. It is the resolution of this ignorance for the special case of the Stern–Gerlach experiment which gives it its special place in measurement theory.

It may of course be replied to those points that the Stern–Gerlach experiment does not really constitute a completed measurement unless and until the positions of the silver ions after their passage through the magnet have been registered—in some observer's consciousness even. And in this way the collapse of the wave-packet can be saved by introducing a stage into the measurement process of which we are ignorant (and during which we cannot therefore rule out the occurrence of a collapse of the wave-packet.)[28] But this view seems implausible.[29] Surely the Stern–Gerlach experiment constitutes a measurement just in so far as it correlates a silver ion's possible y-component spin values with macroscopically distinguishable z-displacements. Whether those macroscopic traces, in the form of z-displacements, are then actually observed, let alone taken into consciousness, is surely irrelevant. And if to deny this is the price of saving the collapse of the wave-packet then surely that is too high a price to pay, always given of course that we can do without the collapse in other contexts—as I shall argue below.

So it seems that the third argument we gave in favour of the hypothesis of the collapse of the wave-packet can be undermined by the fact that it is inconsistent with the laws of QT. It is this paradoxical situation of conflicting reasons both for and against the collapse of the wave-packet hypothesis which has moved Cartwright (1983) to forgo her earlier commitment (Cartwright (1974)) to realism about QT in favour of an instrumentalist attitude: 'In quantum machanics in particular I think there is no hope of bringing the phenomena all under the single rule of the one Schrödinger equation. Reduction of

---

[28] London and Bauer argued this way—see their (1983) p. 252.
[29] On this point see Wigner (1967) pp. 160 ff.

the wave-packet occurs as well, and in no systematic or uniform way . . . the realist programme that I defended in 1974 fails' (p. 165).

I propose to resolve this paradoxical situation differently—more in keeping with Cartwright's earlier realism of (1974), and against taking seriously the collapse of the wave-packet. In particular in Chapter 5 I shall present a resolution of Schrödinger's cat paradox which does not assume a collapse of the wave-packet. And in the next section of this chapter I shall explain why the result of repeat measurements are correlated, without assuming a collapse of the wave-packet, viz. by appealing to the particle aspect of q-systems. These last two moves mean that the only reason left for favouring a collapse of the wave-packet lies in the rhetorical practices of physicists—their description of Stern–Gerlach experiments as *state preparations*. But surely this is a poor reason taken by itself, i.e. if all other reasons for proposing a collapse of the wave-packet can be undermined, then surely the correct course of action is to revise this rhetoric on the grounds that (as I have shown) it contradicts the laws of QT.

This repudiation of the collapse of the wave-packet has removed the threat which was posed to the interpretation of $f(S, t)$ as a field-function by the argument at the beginning of this section, i.e. by repudiating the collapse we no longer face the superluminal travel of the edge of the field described by $f(S, t)$.

We are now also in a position to justify the claim made earlier that the region over which the field associated with a q-system is spread may not be the same as the region over which its location is in-determinate (and which we take as its extension). The justification of this claim requires the introduction of a rule which I shall be justifying in detail in Chapter 4. That rule says that the q-quantity Q for S has a determinate value at t if the density operator for S at t is diagonal in the eigenvectors of Q after it has been projected onto the non-decomposable Hilbert space for S on which Q is defined. From this it immediately follows that for each Stern–Gerlach silver ion e at t', when it emerges from the magnetic field, it is determinate what its coarse-grained z-displacement is, i.e. it is determinate whether it is in the positive or negative half of the z-plane. Also its y-spin at t' is determinate. This is because the density operator for e at t projected onto the (non-decomposable) spin space is:

$$|c_1|^2 \hat{P}(f_{y, 1/2}) + |c_2|^2 \hat{P}(f_{y, -1/2}),$$

and the density operator for e at t' projected onto the space on which the position of e is defined is:

$$|c_1|^2 \hat{P}(f_{z,+}) + |c_2|^2 \hat{P}(f_{z,-}),$$

which are both diagonal in the relevant eigenvectors. (Note that this is despite the density operator for e at t' being given by the projector onto (i), and *not* by (ii).)

Thus any one Stern–Gerlach silver ion has a perfectly determinate coarse-grained z-displacement—its z-displacement either determinately positive or negative—and yet its associated field is spread over all space and in particular over locations with both positive and negative z-coordinates. Thus we trivially see that the region of indeterminacy for each such silver ion does not coincide with the spread of its pilot-field. (Note that this same possibility did not arise for the double slit electrons because their states do not exhibit the same entanglement between position eigenvectors and their spin eigenvectors.)

Finally in this section I note that the latter views provide yet further argument, if such were still needed, for the reality of the probability field described by the state-vector. In the Stern–Gerlach experiment each silver ion's probability field is spilt into two halves by the magnetic field: one half in the positive region of the z-plane, the other half in the negative region. Moreover, on the interpretation I have been suggesting here, each silver ion is determinately in only *one* of these two regions. Therefore part of each silver ion's probability field is 'unaccompanied': with no silver ion in its core. Yet the existence of each half of the probability field is physically significant, i.e. has physical effects. In particular this is apparent if we interact the two halves of the probability field by passing the silver ion and its pilot-field through a second magnetic field anti-parallel to the first. In that case what we observe (and what QT predicts) is that the silver ions are each restored to their initial state. In other words, the two field portions for each silver ion interfere constructively with each other to recreate the initial state. Thus the probability field, even when physically separated from the silver ion, still has a physical existence.

3. PARTICLES AND NON-LOCALITY

In this section I shall first consider an argument in favour of the idea of q-systems as particles, or at least having in part a particle-like nature, and in particular against the identification of q-systems as fields. Consider the following feature of the probability field for the q-system S at time t. For every point x in its domain D there is a propensity for a detection of S in the neighbourhood of x to have a positive result at time t. Moreover there is a correlation between the results of these various detections such that a positive detection of S at location x at t means that any simultaneous attempt to detect S at some other location x' has a negative result. (Grangier, Roger, and Aspect (1986) have performed such correlation experiments on photons.) It is in order to explain this correlation that we think of S as having a particle nature so that it is literally at only one of the locations in D at t, or at least (allowing for S to be a fuzzy particle) manifests itself at only one such location. Indeed once we give up the idea of S as particle-like, by adopting the view of q-systems as fields, then this correlation becomes quite mysterious. In short it seems that we ought to adopt a model for q-systems as particles, or at least as having in part the nature of a particle; but in neither case take them as identical with their probability fields.

A particle model also helps us to explain the perfect correlation between the results of a Stern–Gerlach measurement of y-spin and an immediately subsequent measurement of y-spin on the same silver ions, without assuming a collapse of the wave-packet (as promised in the last section). It explains this *explanandum* by using the result introduced at the end of the last section which tells us that each Stern–Gerlach silver ion at t', after emerging from the magnetic field, has a determinate coarse-grained z-displacement (either positive or negative) and has a determinate y-spin (either $1/2$ or $-1/2$). By using this result we transform the latter *explanandum* into the question of why a y-spin measurement which starts at t', on those silver ions which (determinately) have positive z-displacement, always yields the measured value $1/2$. The answer to this question is simply that there is a perfect correlation between each such silver ion having y-spin $1/2$ and having positive z-displacement at t'. This answer would of course be quite implausible if we saw the silver ions as identical with and hence coextensive with their probability fields, because then there would be no sense in singling out for any one silver ions a half of the z-plane, either the positive or the negative half, as that half in which it is determinately located.

This answer does however raise some difficulties (difficulties we shall return to in Chapter 5, but which we shall introduce here). First it indicates that there are two quite different possible states, one pure with the state-vector (i) (of section 2) and the other mixed with the density operator (ii) (of section 2), which exhibit the same correlation, viz. positive and negative z-displacements correlated with y-spin values $1/2$ and $-1/2$ respectively. But whereas the existence of the latter correlation between possessed values is clear for states defined by the density operator (ii) (as we shall see), it is not so clear for the pure state (i). Indeed we can justify the existence of the latter correlation for the pure state (i) only by working back (deriving it) from the very correlation between measured values which is our *explanandum*. And this means that our explanation here takes on an appearance of circularity, i.e. the *explanans*, which includes the former correlation between possessed values, is justified by the very same proposition, viz. the correlation between measured values, which we are trying to explain (the *explanandum*). But this sort of circularity is quite acceptable in the context of explanations, if not predictions. (Cf. Scriven's example of the collapsing bridge in his (1962)). And so the circularity which threatens to engulf us here can, it seems, be rendered harmless. (This will be discussed in section 4, Chapter 5.)

We can now combine all the arguments so far, and show that what I called 'the modified pilot-field interpretation of q-systems' emerges as the preferred interpretation. . . Drawing on the arguments of the previous chapter, we can object to Bohr's interpretation on the grounds that it does not interpret the state-vector as describing an always present probability field. And more generally we can object to it on the basis that it too readily gives up the entrenched metaphysical idea of a unified (temporally invariant) model for q-systems. Moreover, as we argued in section 1 of this chapter, it does not correctly locate the explanatory principle for interference effects, i.e. it takes the explanatory principle to lie with the indeterminacy in position of q-systems, whereas in fact it lies with the wave form of the probability field. And this last objection, as we saw in section 1, also works against Born's unmodified pilot-field interpretation. On the other hand in this section we have argued against the view which simply identifies S at t with the probability field $f(S, t)$. Moreover in the last section we provided the materials to complete the argument against the 'dual aspect' (hybrid) model for q-systems—as permanent hybrids of fields and particles—which was foreshadowed in section 1, where we pointed out that the dual-aspect model could not cope with the

Stern–Gerlach experiment, because it could not explain how the probability field associated with each silver ion, but not the silver ions themselves, were split by the magnetic field. And that leaves the pilot-field interpretation as the only candidate holding the field.

It is perhaps appropriate here briefly to comment on what happens to the above arguments in the context of Quantum Field Theory. In that case it is more plausible to take the probability fields (now quantized in their own right) as fundamental; with particles emerging as epiphenomena of the quantization of various quantities associated with the field (such as energy and momentum). In that case questions such as whether the same particle is registered by several different measurements at a given time cease to be of interest, as long as the relevant 'occupation numbers' (how many particles of which type are registered in the various energy/momentum states) satisfy the relevant conservation laws for charge, energy, momentum, etc. which are generated by the Lagrangean's invariance properties. But discussion of this takes us beyond the issues which we can discuss here.

In sum what picture do we now have of the Stern–Gerlach experiment? Are we to picture the splitting of the incident beam of silver ions by the magnetic field on the model of each silver ion totally ending up in one or other of two secondary beams of reduced intensity, or do we rather picture each silver ion as somehow divided into two 'ghost' silver ions which may then interfere with each other? The answer is 'Neither'. The first picture is accurate if we focus just on the position of the silver ions; but it also implies a collapse of the wave-packet, and hence its inadequacies become apparent once we consider more complex quantities such as $s_z \times p_z$. The second picture is also shown to be inadequate by the evident fact that each silver ion has either a positive z-displacement or a negative z-displacement at t', not both (we observe this by allowing the silver ions to fall one by one on a photographic plate). Instead I have suggested that we see each silver ion as a particle piloted by its own probability field. This field is split into two halves by passing the silver ion between the poles of a magnet but the silver ion itself is not split. Indeed each silver ion is either deflected in the positive or negative z-direction, thereby becoming separated in space from one-half of its probability field. The two halves of this field still interact (interfere) with each other however, thereby affecting the path of the silver ion (and in particular producing the interference effects characteristic of the equation (i) as opposed to (ii) of the last section). However, as Born emphasized, the field here must not be seen as carrying energy or momentum—or mass or charge for that matter—*contra* Schrödinger.[30]

In the following chapters I shall extend this modified pilot-field interpretation for q-systems by entering into a discussion of mixed states, treating them as states *sui generis*, and not just as pure states of which we happen to have incomplete information. This will necessitate yet a further abstraction of the notion of the pilot-field: from being represented by a state-vector to representation by a density operator (an operator on vectors). This extension will also enable a further critique of the orthodox interpretation by an extension of the rather restrictive Bohrian criterion for q-quantities to have determinate values. Despite all these extensions and modifications however, what emerges never strays too far from the orthodox interpretation, particularly in its later (post-Bohr) phase. In particular the later Heisenberg's concept of a q-system as a node of potentia, described by a state-vector[31] (or even density operator), is not a world apart from the concept I shall be endorsing here of q-systems as non-classical fuzzy particles accompanied by (steered by) probability fields represented by density operators.

There is a final important point to make in connection with the correlation cited at the beginning of this section. The Reichenbach principle of prior common cause says that if A and B are positively correlated then either they are directly causally related (A is a cause of B or vice versa) *or* there is a prior common cause of A and B.[32] But clearly there cannot be a prior common cause for the detection of S at t at x and the non-detection of S at t at x', because, we suppose, there is no prior cause for either of these detections taken individually, i.e. we suppose the detections are both strongly indeterministic in that there are no prior causes for their results. Hence by the Reichenbach principle it must be the detection of S at t at x which simultaneously causes the non-detection of S at t at x' (or vice versa). But this would contradict the well-entrenched principle of local causation—that one event cannot affect another event which occurs at a distance from but simultaneously with the first. Thus it seems that without making any assumptions beyond (a) the existence of the above correlations and (b)

---

[30] See Born's 1955 letter to Renninger quoted in Jammer (1971) p. 495, which says: '[the] waves have some sort of reality, but . . . are not carriers of energy or momentum'.

[31] Heisenberg (1971) p. 160.

[32] See, for example, Salmon (1975) pp. 121 ff.

the acausal nature of the individual detections, i.e. assuming no prior causes for the individual detections, we have derived a causal anomaly, viz. instantaneous action at a distance. How can QT be saved from this conclusion?

One could perhaps object to assumption (b), i.e. to the assumed acausal nature of the detection process. However objecting to (b) turns out to be of no use for two reasons. First the assumption (b) is bypassed in the more sophisticated version of the above argument given by Bell (1981), and which will be referred to in Chapter 8. In Bell's version of the argument possible prior causes for the outcomes of detections are allowed, albeit restricted to the values taken by hidden variables; and he then eliminates the possibility of the two detections having a prior common cause by showing that such a possibility implies the Clauser–Holt–Horne–Shimony inequality which not only contradicts QT but is experimentally refuted. This only leaves the possibility of the one detection directly affecting the outcome of the other detection—and hence instantaneous action at a distance, i.e. non-locality.

Objecting to assumption (b) turns out to be of no help for another reason too, viz. that the above form of argument leads to objectionable conclusions even in situations where (b) is not at issue. For example consider a classical point particle S in Brownian motion, changing its position acausally at a sequence of times $t_1, t_2, \ldots$, so that there is a non-trivial probability distribution over a range of possible positions for S at any time after t_n, and the characteristics of S (or its environment) at any t before t_n have no influence on its position after t_n, for any n. But clearly at time t, just after t_1, the following holds for any $x \neq x'$:

Prob $(S$ is at x' at t/S is at x at $t) = 0$,

or equivalently:

Prob $(S$ is not at x' at t/S is at x at $t) = 1$.

And since *ex hypothesi* Prob $(S$ is not at x' at $t) < 1$ we see that S not being at x' is positively correlated with S being at x at t. And, by following exactly the reasoning given above, we derive that S being at x at t causes S not being at x' at t or vice versa, and hence we again have a causal anomaly.

Hellman discusses a similar argument to this in his (1982) p. 467, but diagnoses the problem with it to be that 'the fundamental physical probabilities have been conflated with ordinary epistemic ones'. I draw a different conclusion about what has gone wrong. In particular I conclude that the Reichenbach prior common cause principle simply does not apply where the correlation is between the outcomes of acausal processes. In short I deny that there is any causal connection between S being at x at t and S not being at x' at t. This denial provides an immediate counter-example to Reichenbach's principle, but also of course saves us from having to postulate a causal anomaly in the above situation. How then do I explain the above correlation if not by a causal connection? It is simply by S being a particle with a unique position at time t. There is no need to invoke any causal principle in addition to this in order to explain the correlation.

My rejection of the Reichenbach principle here also explains why in a later chapter I focus on Stapp's and Eberhard's versions of Bell's proof that 'realism' in QT implies causal anomalies of the sort described above, viz. instantaneous action at a distance. I prefer these versions to Bell's own version[33] because in his version, unlike Stapp's or Eberhard's, the causal anomaly is derived from a positive correlation between the results of spatially separated measurements via the implicit use of Reichenbach's principle. But it is precisely the use of Reichenbach's principle in this context which is in doubt, for the reasons given above, viz. that the measurement processes are acausal. (This is not of course to deny that Bell's proof is useful in ruling out a wide range of stochastic local hidden variables theories. What I do deny is that Bell's proof shows that local realism is impossible, and I deny this because Bell's proof makes use of the invalid Reichenbach principle.)

In summary then we see that the apparent causal anomalies which threaten QT in virtue of the above correlations can be explained away by arguing that the Reichenbach common cause principle does not apply to the relevant situations because of the failure of determinism.

[33] See Bell (1981).

4

Operators and Physical Quantities

Vectors; state-functions and state-vectors; operators and physical quantities; density operators; the ignorance interpretation of mixtures; superselection rules.

1. DENSITY OPERATORS AND MIXTURES

I shall start this chapter by summarizing some of the claims of Chapter 2. In QT we represent the states of systems in QT by vectors. In particular with each q-system S is associated a set (or space) $\mathcal{H}(S)$ of vectors each one of which represents a possible state of S at t. Each vector in $\mathcal{H}(S)$ is represented by a set of N numbers $(c_1, c_2, \ldots, c_N)$. The number N is called 'the dimension of $\mathcal{H}(S)$', and the various c_n for $n = 1, 2, \ldots, N$ are called 'the vector components'.

For the purposes of QT the important relation between any pair of vectors is their 'scalar product', where the scalar product of f and g is defined as follows: if f, g are any two vectors, represented by their sets of components $(c_1, c_2, \ldots, c_N)$ and $(d_1, d_2, \ldots, d_N)$ respectively then the scalar product of f and g—denoted '(f, g)'—is

$$c_1 d_1 + c_2 d_2 + \ldots + c_N d_N,$$

which is abbreviated to:

$$\sum_{n=1}^{N} c_n d_n \text{ or just } \Sigma c_n d_n.$$

For simplicity in this chapter I assume that we deal with 'real vectors', i.e. vectors with real numbers as components, and in a finite dimensional $\mathcal{H}(S)$, i.e. N finite.

Note that the scalar product of two vectors must be distinguished from the product of a vector with a number. Thus the product cf, of the number c with the vector f, where f is the vector $(c_1, c_2, \ldots, c_N)$, is simply defined as the vector $(cc_1, cc_2, \ldots)$. The sum of vectors is defined in a similarly obvious way (with the vector sum $f + g$ having components $(c_1 + d_1, c_2 + d_2, \ldots)$). A linear combination of vectors $f_1, f_2, \ldots$ is any vector of the form $c_1 f_1 + c_2 f_2 + \ldots$ where $c_1, c_2, \ldots$ are numbers.

The scalar product of two vectors features in the bridge principle which connects the vector representation of a system's state with the so called observables. This bridge principle is called 'the Born interpretation'. It supposes that with each discrete valued physical quantity Q for S is associated a set of 'eigenvectors' in $\mathscr{H}(S)$, one for each possible value q of Q if the physical quantity is non-degenerate. The eigenvector $f(Q, q)$ of the non-degenerate Q for possible value q is defined as the vector representing that (unique) possible state of S for which if S is in that state at time t then $P(Q, q, S, t) = 1$. The Born interpretation tells us that in general $P(Q, q, S, t) = (f(Q, q), f(S, t))^2$ where $f(S, t)$ is the vector representing the state of S at t. This Born interpretation is extended to the more general case of degenerate physical quantities (for which there may be several non-trivially distinct eigenvectors for each possible value) in Appendix 3, and is extended to cope with continuous-valued physical quantities in Appendix 4.

In some contexts it is as a vector that we represent the states of q-systems, but in other contexts it is functions—indeed functions of position—which are used to represent the states of q-systems, i.e. instead of there being a vector associated with a q-system at any given time to represent its state there is a function of position—a 'state-function'. The bridge principle which connects the state-function $f(S, t)$ of S at t with observables tells us that $f(S, t)(x)$—the value of $f(S, t)$ at location x—satisfies the relation:

$$|(f(S, t)(x))|^2 \, dx = P(X(x, dx), S, t),$$

where the right-hand side probability is the probability of S at t being measured to have a location within the interval of position values from x to $x + dx$. (For simplicity, I suppose the space in which S at t is located is just one-dimensional.) In other words $f(S, t)(x)$ gives us an estimate of how likely it is that S would appear in the neighbourhood of the point x at time t if we were to look for it there.

How are we to reconcile these two methods of state representation —one in terms of vectors, the other in terms of functions of position? Very simply by pointing out that a suitable set of functions of position may be regarded as an instance of an abstract vector space: they satisfy all the formal requirements we make of abstract vectors. In particular suppose we restrict ourselves to functions of positions f which can be written as a linear combination of some 'basis' set of functions $f_1, f_2, \ldots, f_N$, i.e. for each f there is a unique set of numbers $c_1, c_2, \ldots, c_N$ for which $f = \Sigma c_n f_n$, where the sum $f_1 + f_2$ of two functions f_1, f_2 is defined in the usual way:

$$(f_1 + f_2)(x) = f_1(x) + f_2(x),$$

and so too is a linear multiple cf of a function f, i.e. $(cf)(x) = cf(x)$. Then each such f is uniquely characterizable by the set of numbers $(c_1, c_2, \ldots, c_N)$, just as vectors are. Moreover a scalar product is definable between such functions as above. Thus we see that the vector representation of states can be taken simply as an abstraction from the function-of-position representation, an abstraction in which certain irrelevant features of the latter representation are eliminated.

We also need to introduce the idea of an operator on a vector space. This is partly to facilitate the introduction of the density operator, but also to reconcile the treatment of physical quantities in QT given here with the more usual treatment (in which q-quantities are represented by operators). An operator on a vector space is simply defined as an operation which turns vectors into other vectors. For example an operator $\hat{A}$ may be defined as follows:

$$\hat{A}(c_1, c_2, \ldots, c_N) = (kc_1, kc_2, \ldots, kc_N).$$

This operation $\hat{A}$ is the operation of multiplying by a constant k, as defined on the N-dimensional vector space $\mathscr{H}_N$.

Operators are important in traditional treatments of QT in that they are used to represent physical quantities. How does this form of representation connect with the form of representation which I have been using, viz. to characterize physical quantities in terms of their eigenvectors and possible values? The answer to this question is simple in principle, although the details become complicated in practice.

Let $\mathscr{H}(S)$ be the Hilbert space of possible states for S. Then any physical quantity Q which is a physical quantity for S is represented by that operator $\hat{Q}$ (an operator on $\mathcal{H}(S)$) for which the 'eigenvalue equation':

$$\hat{Q}f = qf$$

is satisfied for all those and only those numbers q and vectors f in $\mathcal{H}(S)$ for which q is a possible value of Q and f is an eigenvector of Q for possible value q. Thus to specify $\hat{Q}$ is to specify the possible values and eigenvectors of Q in $\mathcal{H}(S)$, and hence to represent Q. This operator representation is discussed in more detail in Appendix 1.

The operator representation of physical quantities, although conceptually more complex, has certain heuristic advantages. This is because the algebraic relations between operators representing physical quantities in QT are isomorphic with the algebraic relations between the corresponding representatives of those quantities in classical mechanics. This isomorphism provides a rich source of information about the representation of physical quantities in QT.[1] Also, as we see in Appendix 2, the operator representation suggests a natural way of extending the treatment of discrete-valued physical quantities within QT to apply to continuous-valued physical quantities.

A particularly important operator for the purposes of QT is the projection operator onto a vector. This is defined as follows:

(Defn) $\hat{P}(f)$, the projection operator onto the vector f in $\mathcal{H}(S)$, is the operator which maps any vector g in $\mathcal{H}(S)$ onto $(f, g)f$, i.e. $\hat{P}(f)g = (f, g)f$.

(Note $(f, g)g$ is the product of the vector g with the number (f, g), and hence is itself a vector, as indicated above.)

We can now introduce the density operator. This is as part of a generalization of the notion of a state in QT. Hitherto we have associated a unique vector $f(S, t)$ with each q-system S and time t, and defined it by its role in the equation:

(BSI) $P(Q, i, S, t) = |(f(Q, i), f(S, t))|^2,$

where $f(Q, i)$ is the unique eigenvector of non-degenerate Q for possible value q_i. Those S and t for which such an $f(S, t)$ exists will from now on be regarded only as special cases, and if such an $f(S, t)$

---

[1] See, for example, Dirac's treatment (1958) of the formal isomorphism between commutators and Poisson brackets.

exists then S at t will be said to be in a pure state. There is however a more general class of states than pure states. In particular S at t is said to be in a 'mixed state' iff there is a set of mutually orthogonal vectors $f_1, f_2, \ldots$ and associated probability coefficients $p_1, p_2, \ldots$ where $\Sigma p_n = 1$, such that for any Q, i,

(i) $\quad P(Q, i, S, t) = \sum_n p_n |(f(Q, i), f_n)|^2.$

I denote the mixed state for S at t in which (i) is satisfied for all Q, i by '$\{p_n, f_n\}$'. (In the special case where all the f_n are identical to within a constant phase factor, or $p_n = 1$ for some n, this mixed state becomes a pure state with state-vector f_n to within a constant phase factor.) Gleason's theorem tells us that more general states than this are not possible if we insist on preserving the condition (exemplified in (BSI)) that for non-degenerate $Q, P(Q, i, S, t)$ is a function of $f(Q, i)$ for given S, t.[2]

Now suppose we broaden the class of q-quantities to include degenerate q-quantities. We know (see Appendix 3) that the $P(Q, i, S, t)$, for S at t in the pure state $f(S, t)$, are then given by:

(BSI)' $\quad P(Q, i, S, t) = \sum_x |(f_{ix}, f(S, t))|^2,$

where $\{f_{i1}, f_{i2}, \ldots, f_{ix}, \ldots\}$ is any basis set in V_i, the closed linear subspace of eigenvectors of Q for value q_i.[3] If we now insist on the condition (exemplified in (BSI)') that $P(Q, i, S, t)$ is a function of V_i for given S, t (or, equivalently, that $P(Q, i, S, t)$ is a function of $\hat{P}(Q, i)$ the projector onto V_i), then Gleason's theorem tells us that the most general possible form for all the $P(Q, i, S, t)$ for given S, t is that there are $\{p_n, f_n\}$ for which

(BSI)* $\quad P(Q, i, S, t) = \sum_n p_n \sum_x |(f_{ix}, f_n)|^2$ for all Q, i.

If the $P(Q, i, S, t)$ take this form then S at t is said to be in the mixed state $\{p_n, f_n\}$.

Prima facie there seem to be two different ways to obtain mixed states, as so far characterized. One way it seems is for S at t to be in the

---

[2] Gleason (1957).
[3] A closed linear subspace of vectors is any set of vectors which is closed under linear combinations of its members (and also contains the limits of such linear combinations, if they exist).

mixed state $\{p_n, f_n\}$ as a result of S at t being produced by one of a set of different possible 'state preparations' $M_1, M_2, \ldots, M_n \ldots$, where, for each n, there is a probability p_n of M_n occurring, and M_n produces S at t in a pure state with state-vector f_n. There is however a problem in making sense of this suggestion. Even though there is probability p_n of M_n taking place, for each n, where none of the $p_n = 1$, nevertheless *one* of the M_n does take place—let it be M_1. As such S at t really has the state-vector f_1, i.e. is in the pure state f_1, so that

(i)' $P(Q, i, S, t) = |(f(Q, i), f_1)|^2.$

But this contradicts (i). And this same contradiction emerges if we take S at t as the terminus of an indeterministic process which produces outcome f_n with probability p_n for all n.

We can avoid this contradiction however, by interpreting the p_n and hence the $P(Q, i, S, t)$ as epistemic, i.e. we claim that if we know that f_1 is the state which S at t is in (because, for example, we know that it was M_1 which prepared S at t) then of course (i)' holds. But this does not contradict (i) because the p_1 referred to in (i) becomes 1 precisely if we know that S at t is in f_1. Indeed it is only because we do not know which of $f_1, f_2 \ldots$ characterizes S at t that $p_n < 1$ for all n. (The subjective nature of the probabilities here can be papered over, if not removed, by estimating them in terms of corresponding relative frequencies over ensembles, via the 'statistical syllogism'.[4] In particular we estimate $P(Q, i, S, t)$ by the relative frequency with which measurements of Q in S register the value q_i in situations for which S is characterized by the same facts that we know apply to S at t.)

Alternatively we can remove the above contradiction by rejecting the states of QT as properties of single systems, and taking them instead as properties of ensembles. Thus we see an ensemble E of systems as in the mixed state $\{p_n, f_n\}$ at t iff E is composed of various sub-ensembles E_n in the pure state f_n with p_n as the fraction of E constituted by E_n. We then simply reject as improper any question of whether a particular member of E_n at t is really in the pure state f_n or the mixed state $\{p_n, f_n\}$. This question is improper because mixed and pure states are states of ensembles not of single systems. The latter view is taken by d'Espagnat for example,[5] and was also taken by this author in his earlier publications.[6]

---

[4] See van Frassen (1980), ch. 6, sect. 2.2.
[5] D'Espagnat (1971), ch. 6.   [6] e.g. see Krips (1969).

But QT allows yet another way to produce S at t in a mixed state $\{p_n, f_n\}$. That way is to prepare S as a component of some combined system $S + S'$. The point is that we can prepare $S + S'$ to be in a pure state F at t but S may nevertheless be in a mixed state at t in the sense that there are non-trivial $\{p_n, f_n\}$ for which (i) holds for all Q, i, and moreover there is no vector f in $\mathcal{H}(S)$ for which S is in f at t. (This is shown at the end of this chapter, via a theorem from Appendix 6.) In other words to single out a state-vector for $S + S'$ does not always single out a state-vector for S. This phenomenon appears to be quite non-classical, i.e. classically we expect the highest-grade information about $S + S'$ to imply highest-grade information about S. But this fails in QT *if* we see the specification of a mixed state for S at t as 'lower-grade information' than a specification of a unique state-vector for S at t. The appearance of paradox here can however be removed by pointing out that to see the specification of a mixed state for S at t as somehow less informative than the specification of a state-vector is to beg the question because it seems to presume that S is always 'really' in a pure state.[7]

In this book I shall only refer to these last sorts of mixed states as 'mixed states', i.e. I shall use the term 'mixed state', or 'mixture', to refer only to states of sub-systems of systems which are themselves in pure states. As such 'pure' and 'mixed' become *exclusive* and *exhaustive* qualifiers of states in QT (except in the special case of a mixture $\{p_n, f_n\}$ where $p_n = 1$ for some n or the f_n are all the same to within a phase factor). This stipulation means that the probabilities p_n which define the mixed state $\{p_n, f_n\}$ for S at t are 'irreducible', i.e. we can change them only by changing the state-vector of some system $S + S'$ at t. This stipulation reverses the implication behind the (for my purposes) unfortunate convention which d'Espagnat adopts.[8] For d'Espagnat mixtures of the sort which I take to be mixtures are termed 'improper', whereas mixtures in which the probabilities are reducible are termed 'proper'. The justification for reversing d'Espagnat's convention is that (a) it restores the idea that pure states and (proper) mixed states are exclusive (except in the trivial case where $p_n = 1$ for some n or all the f_n are identical to within a phase factor); (b) it eliminates subjective state descriptions from among the

---

[7] On the other hand we can make *formal* sense of the claim that to specify $\{p_n, f_n\}$ is 'less informative' for non-trivial $\{p_n\}$ than for $p_n = 1$ for some n, i.e. we simply define information as negentropy: $\Sigma\, p_n \log p_n$, in the usual way.

[8] D'Espagnat (1971), ch. 6, sect. 3.

'proper' descriptions allowed by QT (by eliminating epistemic probabilities); and (c) it restores the idea of pure and (proper) mixed states being states of single systems rather than ensembles.

This stipulation also represents a total rejection of the 'ignorance interpretation of mixtures',[9] according to which the p_n for any mixed state $\{p_n, f_n\}$ are mere epistemic probabilities. But, I shall argue, this rejection is no great loss, because the latter interpretation is beset with difficulties (discussed by van Frassen for example)[10] some of which I shall rehearse below. In particular I shall show that the ignorance interpretation of mixtures contradicts certain plausible principles of QT. Indeed this contradiction is the main reason I have for rejecting the ignorance interpretation, and substituting for it the stipulation that it is only what d'Espagnat terms 'improper mixtures' which are mixtures (*sic*).

I also incorporate the following principle into QT:

(Princ)   If $P(Q, i, S, t) = P(Q, i, S, t')$ for all Q, i then the same q-quantities have determinate values in S at t, with the same probabilities, as in S at t'.

The latter principle is one which both Einstein and Bohr would accept, although for Einstein it is a trivial consequence of his 'realist' principle that all physical quantities have determinate values *and* that the probability $P(Q, i, S, t)$ of measuring Q to have a value q_i in S at t is equal to the probability of Q determinately having that value in S at t. On the other hand, for Bohr, (Princ) is justified quite differently. For Bohr the $P(Q, i, S, t)$ determine the state of a system (as represented by its state-vector say), and the state of a system then uniquely determines which, if any, physical quantities have determinate values (and with what probabilities). Thus Bohr would also accept (Princ) even though he would not agree with Einstein that the $P(Q, i, S, t)$ simply reflect the probabilities of physical quantities actually having various possible values.

The rationale for this (Princ) will be clear to the positivistically minded. The $P(Q, i, S, t)$ are the only directly observable manifestations of the probability distributions over possible values for physical quantities in S at t, and hence if $P(Q, i, S, t) = P(Q, i, S, t')$ for all Q, i then the positivist would insist that there is no empirically

[9] Put forward by Reichenbach in 1948. See van Frassen's discussion in (1980) p. 171.
[10] See van Frassen (1980) p. 171 and (1973) p. 97.

meaningful distinction to be made between the probabilities of physical quantities having particular values in S at t and the corresponding probabilities for S at t'. Consistent with my anti-positivist stance here, I cannot present this rationale for (Princ). Rather my rationale will be the overall coherence, explanatory fruitfulness, etc. of the version of QT which takes (Princ) as an axiom.

I also assume (with similar rationale) that

(Princ)′   If, for various (but not necessarily all) of the possible values q_n of Q, a *non-degenerate* physical quantity for S, there is probability p_n that Q has value q_n determinately in S at t and $\Sigma p_n = 1$ then S at t is in the mixed state $\{p_n, f_n\}$, where f_n is the eigenvector of Q for value q_n in $\mathscr{H}(S)$ for each n.

(Note that, as we shall show later, the slight generalization of (Princ)′ got by removing the restriction that Q is non-degenerate is invalid.)

It may be thought that a rationale for (Princ)′ can be given as follows. Assume Q has q_n in S at t with probability p_n for all n where $\Sigma p_n = 1$. Then Bayes theorem tells us that

$$P(Q', i, S, t) = \sum_n p_n P(Q', i, S, t/\text{val } Q, n),$$

where $P(Q', i, S, t/\text{val } Q, n)$ is the conditional probability of measuring Q' to have its ith value in S at t, given that Q has the value q_n (in S at t). But the latter probability is the probability of measuring Q' to have its ith value over a sub-ensemble of systems identical to S at t but in which Q always has value q_n determinately. And a sub-ensemble in which Q has value q_n always is surely a sub-ensemble in which Q has value q_n with certainty. But by (BSI) the members of any such sub-ensemble must be characterized by the state-vector f_n where f_n is the eigenvector of Q for value q_n. (Note the assumption that Q is non-degenerate is needed here.) Hence by (BSI) the latter probability is $|(f_i', f_n)|^2$, where f_i' is the eigenvector of Q' for its ith value, Q' being assumed non-degenerate too. Hence

$$P(Q', i, S, t) = \sum_n p_n |(f_i', f_n)|^2,$$

which is tantamount to the relevant 'vanishing of interference effects' between the eigenvectors of Q. Hence, by definition, S at t is in $\{p_n, f_n\}$. QED.

This rationale for (Princ)' is not one I endorse however. First of all the apparently harmless assumption made as part of the rationale that the $P(Q', i, S, t/\text{val}\, Q, n)$ exist, may be questioned (see A. Fine (1979)) as can the frequency interpretation of probabilities which it employs. Moreover (Princ) undermines the above rationale for (Princ)' as follows. The above rationale assumes that a sub-ensemble of systems in which Q has q_n always is an ensemble for which there is probability 1 that Q has value q_n, and hence is in a pure state f_n. This view implies the ignorance view of mixtures however, which as we shall see later is ruled out by the combination of (Princ) and (Princ)'.

Nevertheless, despite my rejection of this rationale for (Princ)', I shall incorporate it into my interpretation of QT. Its justification too will be taken to lie in the overall explanatory power, coherence, etc. of the interpretation in which it plays a role.

From (Princ) and (Princ)' we can now prove that

(Princ)'' If S at t is in a mixed state $\{p_n, f_n\}$, where f_n is an eigenvector of Q for value q_n and $q_n \neq q_m$ for any $n \neq m$ then there is probability p_n that Q has value q_n determinately in S at t for all n.

The proof of this principle for degenerate as well as non-degenerate Q involves yet another plausible principle however (see appendix 3):

(A) Let Q' be non-degenerate in S and let the eigenvector of Q' for value q'_n be the same as an eigenvector of another physical quantity Q for value q_m. Then $P(\text{val}\, Q, m, S, t)$, the probability that Q has value q_m in S at t, $\geq P(\text{val}\, Q', n, S, t)$ for any t.

(The non-technical reader may prefer to skip the next few pages until the paragraph in which the density operator is introduced, at the definition marked '(Defn.)'. The intervening materials provides a proof of (Princ)'', and indicates the surprising but central result that the converse to (Princ)'' is inconsistent with the axioms so far put forward.)

Proof of (Princ)'' (for both degenerate and non-degenerate Q):
Let S at t be in a mixed state $\{p_n, f_n\}$, where f_n is an eigenvector of Q for value q_n and $q_n \neq q_m$ for every $n \neq m$. Let Q' be that non-degenerate physical quantity in S (and I assume that there is such a Q') ior which f_n is the eigenvector of Q' for value q'_n for each n. (The $\{f_n\}$ may not of course be complete, i.e. exhaust *all* the eigenvectors of Q'.)

Now clearly for any Q'', i,

$$P(Q'', i, S, t) = \sum_n p_n (f_i'', f_n)^2,$$

which is just what the $P(Q'', i, S, t)$ should be in the case that there is probability p_n that Q' has value q_n' in S at t for all n (by (Princ)'). Hence there is probability p_n that Q' has value q_n' is S at t for all n (by (Princ)). Hence, by (A), $P(\text{val } Q, n, S, t) \geq p_n$. Now there are two cases to consider: $P(\text{val } Q, n, S, t) > p_n$ for some n or $P(\text{val } Q, n, S, t) = p_n$ for all n. But in the first case $\Sigma_n P(\text{val } Q, n, S, t) > \Sigma p_n = 1$, which contradicts $\Sigma_n P(\text{val } Q, n, S, t) = 1$. Hence $P(\text{val } Q, n, S, t) = p_n$ for all n. Therefore there is probability p_n that Q has value q_n in S at t for all n. QED.

It is important to note that (Princ)' is not the same as the strict converse of (Princ)'', i.e. (Princ)' does not say that if there is probability p_n that Q has value q_n determinately in S at t for all n then S at t is in a mixed state $\{p_n, f_n\}$ for some $\{f_n\}$ such that f_n is an eigenvector of Q on $\mathcal{H}(S)$ for eigenvalue q_n. Indeed, as we shall see, this strict converse of (Princ)'' (which can be seen as a generalisation of (Princ)') is inconsistent with the other principles here. Moreover not only is the strict converse of (Princ)'' inconsistent with the other principles here, but so is the following plausible weakening of that converse:

(Weak Converse) If at t there is probability p_n that Q, a non-degenerate q-quantity for S, has value q_n for all n then S at t is in a mixture of eigenvectors of Q for which the sum of the probabilities associated with eigenvectors for possible value q_n is p_n for all n.

This inconsistency comes about because of the possibility of joint systems in which one and the same physical quantity may count as a physical quantity both for the whole system, in which it is degenerate, and for a sub-system in which it is non-degenerate.[11] Indeed if P^1 is represented by the operator $\hat{P}_1$ on $\mathcal{H}(S_1)$ then it is represented by the degenerate operator $\hat{P}_1 \times \hat{I}_2$ on $\mathcal{H}(S_1 + S_2)$, where $\hat{I}_2$ is the identity operator on $\mathcal{H}(S_2)$. In particular if we let P^1 be non-degenerate on $\mathcal{H}(S_1)$ with g_i^1 as the unique eigenvector of P^1 in $\mathcal{H}(S_1)$ for value p_i for all i, then P^1 is degenerate on $\mathcal{H}(S_1 + S_2)$ with $g_i^1 \times f^2$ an eigenvector of P^1 in $\mathcal{H}(S_1 + S_2)$ for value p_i for any f^2 in $\mathcal{H}(S_2)$. The inconsistency of (Weak Converse) with the other principles given

---

[11] The notions of a joint system, product space, and product of vectors of form $f \times g$, are discussed in Appendix 7.

above can now be shown easily as follows. From (BSI)* as generalized to $S_1 + S_2$ it follows (see theorem in Appendix 6) that if $S_1 + S_2$ is in a pure state $\Sigma c_n f_n^1 \times f_n^2$ where, for each n, f_n^1 is an eigenvector of the non-degenerate physical quantity Q^1 for S_1 for value q_n, and where the set $f_1^2, f_2^2, \ldots f_n^2 \ldots$ is any set of mutually orthogonal vectors in $\mathscr{H}(S_2)$, then

$$P(P^1, p_i, S_1 + S_2, t) = \sum_n |(c_n)|^2 |(f_n^1, g_i^1)|^2,$$

and hence (since P^1 having value p_i is surely the same event whether we consider it in the context of $S_1 + S_2$ or of S_1)

$$P(P^1, p_i, S_1, t) = \sum_n |(c_n)|^2 |(f_n^1, g_i^1)|^2.$$

Moreover clearly S_1 at t is in a (proper) mixed state since there can be no f in $\mathscr{H}(S_1)$ for which f is the state-vector at t of S_1;[12] and hence, by (Princ)″, it follows that there is probability $|(c_n)|^2$ of Q^1 having value q_n in S_1 at t for all n. But for Q^1 to have value q_n in S_1 at t is surely the same as Q^1 having value q_n in $S_1 + S_2$ at t; and hence there is probability $|(c_n)|^2$ of Q^1 having value q_n in $S_1 + S_2$ at t for all n despite $S_1 + S_2$ not being in a mixed state at t, *contra* (Weak Converse).

Note that this same counter-example could have been derived by considering the one system S for which there is a superselection rule, i.e. for which $\mathscr{H}(S)$ is expressible as a direct product $\mathscr{H}_1(S) \times \mathscr{H}_2(S)$ where some physical quantities for S are defined on each of $\mathscr{H}_1(S)$ and $\mathscr{H}_2(S)$. We will consider the extensions of (Princ) and (Princ)′ required by this case in the final section of this chapter. Note also (and this will be critical in our resolution of the Schrödinger cat paradox to be discussed in the next chapter) that not even the following weaker converse to (Princ)″ holds:

(Weaker Converse)   If Q has a determinate value in S at t then S at t is in a mixture of Q eigenvectors.

The same example which we have just considered provides a counter-example to this even weaker converse of (Princ)″ as well.

This (Princ)″ gives special significance to what is called 'the density operator for S at t'. This is defined as follows:

(Defn)   For any S, t, the density operator $\hat{W}(S, t)$, for S at t is $\Sigma p_n \hat{P}(f_n)$ if S at t is in the mixed state $\{p_n, f_n\}$.

---

[12] If there were then the state-vector of $S_1 + S_2$ at t, which *ex hypothesi* exists, would have to be of the form $f \times g$ for some g in $\mathscr{H}(S_2)$—see Appendix 6.

(The sum of operators and multiplication of an operator by a constant are defined in the obvious way, following exactly the corresponding definitions for functions. Indeed operators are functions, albeit from vectors onto other vectors.)

The following theorem is now easily proven (see Appendix 4):

$P(Q, i, S, t) = P(Q, i, S, t')$ for all Q, i iff $\hat{W}(S, t) = \hat{W}(S, t')$.

And the following corollary can then be proven via (Princ)'':

If $\hat{W}(S, t) = \Sigma p_n \hat{P}(f_n)$ where f_n is an eigenvector of Q for value q_n and $q_n \neq q_m$ for all $n \neq m$ then there is probability p_n that Q has value q_n in S at t, for each n.

Note however that the converse to this corollary does not hold, for the same reasons that the converses of (Princ)'' do not hold. Thus we have proven that the density operator of S at t tells us what q-quantities have determinate values in S at t, and in particular we see that the following Bohrian criterion:

(Bohr) Q has determinate value q in S at t if $f(S, t)$ is an eigenvector of Q

can be generalized to mixed states as follows:

(Bohr)'   If $\hat{W}(S, t) = \Sigma p_n \hat{P}(f_n)$ where f_n is an eigenvector of Q for possible value q_n and $q_n \neq q_m$ for all $n \neq m$ then there is probability p_n that Q has the value q_n (determinately) in S at t, for all n.

As indicated above however, the converse to this does not hold even when non-degenerate physical quantities are considered.

The importance of (Bohr)' is that it provides a bridge from observable information about S at t (viz. from information about the $P(Q, i, S, t)$) to information about what values are actually taken by physical quantities in S at t, i.e. to know $\hat{W}(S, t)$ is to be able to say, via (Bohr)', which physical quantities have determinate values and what the probability distribution is over the various possible values. And we may in turn determine what $\hat{W}(S, t)$ is from the various observable $P(Q, i, S, t)$.

The principle (Bohr)' which has been put forward here is also put forward on occasions by both Cartwright and van Frassen.[13] I have

[13] See Cartwright (1974) p. 232. Cartwright uses the term 'sharp value' where I say 'determinate value'. Also see van Frassen's Modal Interpretation in his (1973) p. 100.

derived it however, rather than simply taken it as primitive, as they do. Moreover I have made a point of showing why we must reject its converse given the interpretation of QT which I am advancing here.

(Bohr)′ does however have various prima-facie paradoxical consequences which arise because there may be several (indeed uncountably many) different $\{p_n, f_n\}$ for all of which $\hat{W}(S, t) = \Sigma p_n \hat{P}(f_n)$. This is best illustrated by taking the special case where all the p_n are equal and the set $\{f_n\}$ is a maximal but finite set of mutually orthogonal vectors in $\mathscr{H}(S)$. If the number of various f_n is N then clearly $p_n = 1/N$ (since $\Sigma p_n = 1$). Thus $\hat{W}(S, t) = 1/N \Sigma \hat{P}(f_n)$. But $\Sigma \hat{P}(f_n) = \Sigma \hat{P}(f'_n)$ for any set of vectors $\{f'_n\}$ which are also orthogonal to each other, and maximal in $\mathscr{H}(S)$; and hence for any maximal set of mutually orthogonal vectors $\{f'_n\}$ in $\mathscr{H}(S)$ we have

$$\hat{W}(S, t) = \Sigma p_n \hat{P}(f'_n), \text{ where } p_n = 1/N.$$

This result means that holders of (Bohr)′ are committed to saying that *any* non-degenerate physical quantity Q has some value or other (determinately) in S at t in the above case. But this contradicts what is often taken to be one of the fundamental features of QT, namely the 'indeterminacy principle' that incompatible physical quantities cannot both be assigned determinate values at the same time. (Van Frassen points out a similar difficulty in his (1980) p. 171 and (1973) p. 110.)

I argue that this contradiction is not what it appears to be. In particular I claim that the only restriction on a pair of incompatible physical quantities which QT makes is to preclude both of them simultaneously having a particular value *with certainty*, and in particular that it does not preclude them both simultaneously having a determinate value. In other words the 'indeterminacy principle' as presented above is not part of QT. (I return to this claim in Chapter 5.) Thus we see that there is no contradiction between (Bohr)′ and QT. We can simply accept that any Q has a determinate value in S at t for the special case described above, in consequence of (Bohr)′, and in particular reject the 'indeterminacy principle' as presented above, without prejudicing QT at large.

There are however other difficulties which arise out of (Bohr)′. If incompatible physical quantities can both have particular values in S at t then a joint probability distribution over the various possible values must surely exist. But it is well known from a result by Wigner (1970) that such a distribution cannot exist. A resolution of this difficulty is provided by A. Fine (1974), who argues against assuming the joint probabilities must exist. Also difficulties for (Bohr)′ arise from the Kochen and Specker theorem and Bell's theorem. Those will be discussed in later chapters. (Basically my solution of the latter difficulties will follow van Frassen's de-Ockhamization strategy, and the Redhead–Hellman strategy of rejecting what they call 'determinism'.)

The previous considerations now allow me to expose the main problem for the ignorance interpretation of mixtures, which I alluded to above. Let us suppose, with the ignorance interpretation, that if for all Q', i,

(i) $P(Q', i, S, t) = \sum_n p_n |(f'_i, f_n)|^2$,

then S at t is really in a pure state, viz. one of the f_n. The error in the latter interpretation is now clear when we consider the theorem in Appendix 6 which shows that (i) holds if for some S'

(ii) $f(S + S', t) = \Sigma c_n f_n \times g'_n$,

where the g'_n are orthonormal vectors in $\mathscr{H}(S')$ and $|c_n|^2 = p_n$ for all n. But in that case clearly S is *not* in a pure state with one of the f_n as state-vector, since if it were then (see Appendix 7)

$$\hat{W}(S + S', t) = \hat{P}(f_n) \times \hat{W}(S', t)$$

for some n, which it is easy to show contradicts (ii). Thus we see that the ignorance interpretation of mixtures is inconsistent with certain basic principles of QT.[14] This leaves the interpretation of mixtures which I have been considering, one incorporating (Bohr)′, and stipulating that the only real mixtures are d'Espagnat's improper mixtures, as the only viable candidate for an interpretation of mixtures.

Note that in addition to its crucial role in telling us which q-quantities have determinate values via (Bohr)′, the density operator for S at t also tells us what the various $P(Q, i, S, t)$ are via the Born statistical interpretation in its various forms. In particular, as we see in Appendix 5, we have

(BSI)** $P(Q, i, S, t) = \text{Tr } \hat{W}(S, t) \hat{P}_i$,

---

[14] This same difficulty is pointed out by van Frassen in the appendix to his (1973). Note that the principles used to derive this contradiction do not beg the question by assuming (Bohr)′, but are independently postulated principles of QT.

which is a form for (BSI) general enough to cope with even degenerate Q. (The form of (BSI) for continuous q-quantities is discussed in Appendix 4.)

Where does the field interpretation of q-systems stand after we make the transition to a density operator representation for the states of q-systems? We started with the field for a q-system being represented by a state-function defined over the points in a spatial domain, with the possibility of abstracting from this functional representation to a vectorial representation. In the case of multiple q-systems however, we had to sophisticate this form of representation to allow functional relations *between* space points, rather than functions *on* space points, as the means of state representation. Now it seems that even this sophistication is not enough. It is density operators— operators on functions over space points rather than the functions themselves—which provide the state representation. We are even further from a representation in terms of something like a classical field-function on space points. Nevertheless what we have here is a generalization of the classical notion of a field, in that the existence of several q-systems in a spatial domain is taken to entail the existence *inter alia* of dispositions (propensities) for successful particle detections to take place at various points in the domain (dispositions for which the field is a categorical basis). And these propensities are manifested by the various probabilities for successful particle detections to take place at the various points in the field. Of course there is more to the many-particle field than providing a categorical basis for just these propensities; in particular there are the propensities displayed by the probabilities of success for other sorts of detection experiments, e.g. an experiment for detecting a certain energy. Thus in a broad sense what we have here is a field—a 'probability field' in Born's somewhat misleading terminology—although its form of representation is much more sophisticated than the classical form of representation for fields.

2. SUPERSELECTION RULES

The density operator rules (Princ) and (Princ)', indeed all the theory given in this chapter, was stated on the assumption that there is a unique Hilbert space $\mathcal{H}(S)$ associated with S on which all the relevant q-quantities for S may be defined. Now although this is a valid assumption it is somewhat misleading to leave matters at that. In particular there are systems (such as electrons) which have superselection rules, for which the Hilbert space $\mathcal{H}(S)$ for S is decomposable into a direct product of not further decomposable component spaces:

$$\mathcal{H}(S) = \mathcal{H}(S_1) \times \mathcal{H}(S_2) \times \ \ldots \ ;$$

and some of the q-quantities for S are definable on these component spaces (*as well as* on $\mathcal{H}(S)$ of course). In that case what I suggest is— and this suggestion has already been taken up in the previous chapter—extending (Princ) and (Princ)' to apply to these component spaces in an obvious way. We assume in short that (Princ) applies even if we restrict the Q and q-quantities for S referred to in (Princ) to the q-quantities on a given *component* space for S; and similarly we assume (Princ)' applies if the space for which Q is non-degenerate is only a component space for S (although then of course the $\mathcal{H}(S)$ in (Princ)' must be changed to that component space). If we make these extensions then we are able trivially to adapt the arguments in this chapter to arrive at the density operator rule used in the final section of Chapter 3.

5

Measurement

Ideal measurements; Schrödinger's cat; deriving the Born statistical interpretation; reduction of the wave-packet; Einstein–Podolski–Rosen paradox.

1. DEFINING MEASUREMENTS AND SCHRÖDINGER'S CAT

In this chapter I shall consider particular models for the process of measurement of q-quantities, models which are constructed within QT itself. It is not of course necessary that QT by itself generates such models, any more than we require a description of the functioning of devices for measuring charge, say, to be given entirely from within electromagnetism (although arguably if QT is to be 'complete' then it must provide such models). Indeed Bohr for one seems to have denied precisely this requirement. For him the description of the measurement process for q-quantities ultimately had to be made in classical terms and he seems to have denied that these classical descriptions must in turn be reducible to quantum theoretic descriptions. (The only requirement which Bohr makes is that these two modes of description—the classical and the quantum theoretic—'correspond' to each other in a certain way in their domain of overlap.)[1]

Nevertheless one would expect that QT places certain restrictions on how to measure q-quantities (just as the laws of electromagnetism do play some role in accounts of how to measure charge). These will turn out to be quite strong restrictions, for example fixing the Born

---

[1] In particular he only required an approximate agreement in values between the expectation values for corresponding classical and quantum-theoretic physical quantities; e.g. he required that the expectation value of the quantum theoretic momentum of some collection of particles agrees well with the expectation value of their classical momentum, for situations in which h is small compared to the value of action. See, for example, the discussion of Ehrenfest's theorem in Schiff (1955).

statistical interpretation (BSI) for a certain rather general class of measurement processes. In this way we see (BSI) not as an interpretational principle placed on QT from the 'outside', but rather as emerging from within QT itself. Indeed it is often held that the restrictions placed on the measurement process from within QT are *too* strong in that they impose paradoxical requirements. The Schrödinger cat paradox (1935) and the Einstein–Podolski–Rosen (EPR) 'paradox' (1935)[2] are cases in point. In this chapter I shall be concerned to resolve these paradoxes, by using the interpretation of QT developed in the previous chapters.

I shall use the following 'calibration condition' (Cal) to characterize ideal measurements in QT:

(Cal) An (ideal) measurement of Q in S at t is any interaction between S and a macroscopic system M, which starts at t and terminates at some t' later than t, for which there is a physical quantity $Q(M)$ such that for all i were $f(S, t)$ an eigenvector of Q for value q_i (so that Q has value q_i with certainty) then $Q(M)$ would take its ith value at t' with certainty, where these various values of $Q(M)$ are macroscopically distinct.

For example $Q(M)$ taking its ith value may correspond to the position of some pointer, which is part of M, falling within the ith macroscopically distinct division of some scale which is also part of M. Any M which satisfies (Cal) is called 'a Q-measuring apparatus', and $Q(M)$ taking its ith value at t' is said to be a case of M registering the value q_i for Q at t. (We may also require that $Q(M)$ taking a particular value at the end of the measurement interaction is a 'meta-stable' state of M in the thermodynamic sense; in other words that a measurement involves the creation of a 'trace' or 'mark' in Reichenbach's sense.[3] This requirement means that the observer has a chance to observe what the measuring apparatus registers before what is registered disappears in the course of the natural evolution of the measuring apparatus. This process of achieving meta-stable equilibrium is discussed by Ludwig (1954), Daneri, Loinger, Prosperi (1962), and others, but will not concern us here.)

Note that even if we set aside the previous parenthetical remarks, from a classical point of view it would seem that (Cal) is not sufficient

---

[2] 'Paradox' is placed in quotes here because EPR did not take their argument to constitute a paradox, but merely to demonstrate the incompleteness of QT.

[3] See the discussion in Reichenbach (1956) pp. 198 ff.

to define an ideal measurement. After all even classically we can allow at least the possibility of a measurement which is accurate when the measured physical quantity has a particular value *with certainty* (for each possible value), but may be inaccurate when there is a non-trivial statistical spread over the possible values, and we would not call such a measurement 'ideal'.

For example suppose we (somewhat eccentrically) decide to observe whether a piece of glass breaks on being dropped not by directly observing whether it breaks when dropped but instead by observing the molecular structure of a small splinter of the glass. If the structure we observe is of fragile glass, i.e. glass with a 100 per cent chance of breaking, then we will be able to say that the glass breaks when dropped, and moreover our 'observing' in this way that the glass breaks when dropped will be accurate; and similarly if we observe the structure of the glass to be tough, with a 100 per cent chance of not breaking, then we can be said to have 'observed' (accurately) that it does not break when dropped.

But if the structure is neither of these extremes (tough or fragile), so that we can only say that there is a certain probability less than 1 but more than 0 of it breaking (i.e. it is somewhat fragile and not totally tough), then any conclusion we draw about whether or not it breaks on the basis of this eccentric method of observation runs the risk of error. Thus we have a method of observation of whether or not the glass breaks which works perfectly (*sic*) when the probability of a particular result (either break or not break) is 1, but is less satisfactory if these probabilities are < 1. As such, although (Cal) is satisfied, we would surely refuse the classification 'ideal' for this method of observation. And as such (Cal) is seen to be at best a necessary condition for ideal measurements.

Nevertheless at least in the context of QT, (Cal) has traditionally been taken as definitive of ideal measurements. This is because, as we shall see, (Cal) seems to give the strongest condition which we can guarantee to hold for measurements within the resources of QT. In particular QT offers no means of discriminating between 'really ideal' measurements in which measured values always reflect possessed values say, and measurements which merely satisfy (Cal). I shall return to this point in the next chapter, but for now let me simply say that I shall here follow the tradition of taking (Cal) as definitive of ideal measurements for QT. Moreover for any measurement process which satisfies (Cal) I shall define 'measured value' q.v.:

(Cal)' For all i, Q is measured to have value q_i in S at t iff the Q-measuring apparatus which interacts with S at t ends up by registering the same value which it would register were $f(S, t)$ $= f_i$, where f_i is any eigenvector of Q for value q_i. (This is so whatever $f(S, t)$.)

An immediate corollary of (Cal) is that

(U) If Q and Q' have all their eigenvectors in common, and the corresponding eigenvalues are the same, then a Q-measurement is a Q'-measurement, and vice versa,

Moreover from (Cal)' we see that

(U)' If Q and Q' have all eigenvectors and corresponding eigenvalues in common then for all i, Q is measured to have value q_i in S at t iff Q' is too.

In short we see that (Cal) and (Cal)' mean that physical quantities in QT are measured as functions of their representative operators: same representative operators, same measured values. We will see later that Q and Q', which have all their eigenvectors and eigenvalues in common, and hence have the same representative operators $\hat{Q}$ and $\hat{Q}'$ respectively, may be distinct as physical quantities despite having the same measured values in all circumstances. Indeed, as we shall see in Chapter 7, it is essential that we allow for just this possibility if we adopt the 'realist' principle that physical quantities always have determinate values.

2. SCHRÖDINGER CAT PARADOX

Having defined ideal measurements via (Cal), I shall now show that (Cal) (or even just the uncontroversial weakening of (Cal), which sees (Cal) as merely a necessary condition for ideal measurements) has a paradoxical consequence, the Schrödinger cat paradox.

In QT we associate with any isolated system S' and pair of times t, t' an operator $\hat{U}_{t, t'}(S')$ on $\mathcal{H}(S')$ for which

(i) $f(S', t') = \hat{U}_{t', t}(S') f(S', t)$,

or more generally for which

(i)' $\hat{W}(S', t') = \hat{U}_{t', t}(S') \, \hat{W}(S', t) \, \hat{U}_{t', t}^*(S')$.

This operator is called 'the Schrödinger propagator for S' from t to t''. For our purposes all we need know about it is that it is both linear and unitary in that

$$\hat{U}_{t',t}(S') \Sigma c_i f_i = \Sigma c_i (\hat{U}_{t',t}(S')f_i), \text{ and}$$

$$\hat{U}_{t',t}{}^*(S') \hat{U}_{t',t}(S') = \hat{I}' \text{ the identity operator on } \mathcal{H}(S').$$

Note that implicit in the assumption that this propagator exists is the assumption that isolated systems which are initially in a pure state remain so. This follows from (i)' together with the unitarity of the propagator.[4]

Now suppose $S + M$, a joint system consisting of S and its measuring apparatus M, are isolated so that a propagator $\hat{U}_{t',t}(S + M)$ for $S + M$ exists from time t when the measurement starts to time t' when it terminates. Moreover assume that the measurement interaction conserves the measured physical quantity Q, which is non-degenerate in its form of representation within $\mathcal{H}(S)$. Let $\{f_i\}$ be the eigenvectors of Q in $\mathcal{H}(S)$. Consider a possible case where $f(S, t) = f_i$ and $f(M, t) = F_0$. It then follows (see Appendix 7) that $f(S + M, t) = f_i \times F_0$. And since Q is conserved and non-degenerate in $\mathcal{H}(S)$, it is easy to prove that $f(S + M, t') = f_i \times F_i$ for some F_i and hence $f(M, t') = F_i$.[5] Moreover, by (Cal), we see that F_i must be a state of M in which some physical quantity $Q(M)$, e.g. a pointer-location, takes its ith value with certainty, thereby registering the value q_i. As such F_i is an eigenvector of $Q(M)$ for its ith value, and this is so for all i. Thus we derive the following restriction on $\hat{U}_{t',t}(S + M)$:

(ii) $\hat{U}_{t',t}(S + M) f_i \times F_0 = f_i \times F_i$ for all i where $\{F_i\}$ are eigenvectors of $Q(M)$ for macroscopically distinct values of $Q(M)$.

Note the several assumptions behind this derivation. Firstly there is the assumption that macroscopic physical quantities can be treated as q-quantities, with their own eigenvectors and eigenvalues (an assumption Bohr for one may question—see opening paragraph of

---

[4] i.e. it is easy to show that a unitary transformation of a projection operator preserves idenpotence and hence is still a projection operator: $\hat{U}^*\hat{P}\hat{U} (\hat{U}^*\hat{P}\hat{U}) = \hat{U}^*\hat{P}(\hat{P}\hat{U}) = \hat{U}^*\hat{P}\hat{U}$.

[5] Since $S + M$ at t is in a pure state so must be $S + M$ at t' (given that $S + M$ is isolated). Moreover $\hat{Q}f(S + M, t') = q_i f(S + M, t')$ since Q is conserved by the measurement interaction. But since all eigenvectors of Q in $\mathcal{H}(S + M)$ are of the form $f_i \times F$ for some F in $\mathcal{H}(M)$ it follows that $f(S + M, t')$ is a superposition of such vectors—and hence also of the form $f_i \times F$—let it be $f_i \times F_i$.

this chapter). Next there is the assumption that the possible cases we are considering are 'near enough' to the actual world for all the relevant laws and principles to hold. In this chapter I shall simply take this particular assumption for granted. And it is assumed that both the propagator for $S + M$ and the initial state of M at t are the same in the various possible cases considered as they are in the actual world. In particular it is assumed that the state of the measured system does not place any restrictions on the method of measurement or on the initial state of the measuring apparatus. This is an aspect of what I shall in later chapters call the 'Freedom of Measurement Principle'. All these assumptions are additional to the explicit restriction to a conservative measurement of a non-degenerate q-quantity.

Now consider the possible case for which $f(S, t) = \Sigma c_i f_i$ but $f(M, t)$ is still F_0 and the propagator for $S + M$ is also unchanged. Then

$$f(S + M, t') = \hat{U}_{t', t}(S + M) \Sigma c_i f_i \times F_0$$
$$= \Sigma c_i \hat{U}_{t', t}(S + M) f_i \times F_0 \text{ (since } \hat{U}_{t', t} \text{ is linear).}$$

Hence

(iii) $f(S + M, t') = \Sigma c_i f_i \times F_i$ (by (ii)).

But $f_i \times F_i$ is an eigenvector of $Q(M)$ in $\mathcal{H}(S + M)$; and hence, by (iii), we see that $Q(M)$ is indeterminate in value according to the Bohr (and more generally Copenhagen) criterion that $Q(M)$, a physical quantity for $S + M$, is indeterminate in value at t' if (indeed iff) $f(S + M, t')$ is a non-trivial superposition of eigenstates of $Q(M)$ in $\mathcal{H}(S + M)$ for different eigenvalues. But this contradicts the experimental fact that a measurement interaction always—whatever the initial state of the measured system—terminates with the measuring apparatus determinately registering some value, e.g. with the pointer determinately pointing somewhere on a scale.

This contradiction is illustrated in particularly dramatic form in the Schrödinger cat paradox. Schrödinger (1935) considers a case where the measured Q for S has just the two eigenvectors f_1 and f_2 for eigenvalues q_1 and q_2 respectively, and the measuring apparatus is taken to be a cat in a box with a pellet of poison gas which is broken or not depending on what the state of S at t is. In particular matters are so arranged that the cat 'registers' the value q_1 for Q at t (in the sense of (Cal)) by being dead, i.e. if $f(S, t) = f_1$ then the pellet is broken. And the cat 'registers' the value q_2 for Q at t by being alive, i.e. if $f(S, t) = f_2$ then the pellet is not broken. But clearly in this case whatever the initial state of S it is an experimental fact that M determinately registers some value or other, since cats are always either determinately dead or determinately alive. But it is precisely this conclusion which seems to be violated in the case that $f(S, t) = c_1 f_1 + c_2 f_2$ for $c_1 \neq 0$ and $c_2 \neq 0$.

3. REDUCTION OF THE WAVE-PACKET

The Schrödinger paradox is usually resolved within the Copenhagen interpretation by introducing an extra step in the measurement process, called 'the reduction of the wave-packet' (or what we called 'the collapse of the wave-packet'). We must however distinguish two forms of reduction of the wave-packet, viz. the 'pure-pure form' and the 'pure-mixed' form. The pure-pure form assumes that for a measurement of non-degenerate Q on S at t where $f(S, t) = \Sigma c_i f_i$ (the $\{f_i\}$ being the eigenvectors of Q) and $f(M, t) = F_0$ the measurement interaction continues past the stage when the state of $S + M$ is pure $\Sigma c_i f_i \times F_i$ to a stage when the state of $S + M$ is pure $f_i \times F_i$ for some i. The probability that the final pure state is $f_i \times F_i$ is taken to be $|c_i|^2$, to conform with (BSI).

However some authors put forward a pure-mixed version of the reduction of the wave-packet, which assumes that the final stage of the measurement places $S + M$ in a mixed state characterized by the density operator $\Sigma |c_i|^2 \hat{P}(f_i) \times \hat{P}(F_i)$. Either of these reductions is then taken to justify the conclusion that $Q(M)$ has a determinate value at the end of the measurement, hence solving the Schrödinger cat paradox. (Note that under the ignorance interpretation of mixtures these alternative forms of the reduction of the wave-packet are of course the same, but I shall keep them separate here for the sake of generality. Note also that I shall still let t' denote the time for the termination of the measurement, whether or not the measurement involves this extra reduction process.)

There are however problems in assuming that this reduction takes place (in either its pure-pure or pure-mixed form). For a start it does not conform with the usual way in which states of systems evolve in QT, because it is easy to show that there can be no linear unitary $\hat{U}$ (and hence no Schrödinger propagator) for which, for each choice of $\{c_i\}$ there is some i^* for which

$$\hat{U} \Sigma c_i f_i \times F_i = f_{i^*} \times F_{i^*}.$$

(For proof see Appendix 8.) Nor is there a $\hat{U}$ for which

$$\hat{U} \, \hat{P}(\Sigma \, c_i f_i \times F_i) \, \hat{U}^* = \Sigma |c_i|^2 \, \hat{P}(f_i) \times \hat{P}(F_i).$$

Proof The left-hand side of this last equality is a unitary transformation of a projection operator, and hence is still a projection operator with eigenvalues 1 or 0. But clearly the right-hand side of this equality has eigenvalues $|c_i|^2$ for each i. Thus we have a contradiction except in the degenerate case that $|c_i| = 1$ for some one value of i, and $= 0$ for all other values of 'i'.

Various suggestions have been made in the literature to account for this anomalous reduction. For example London and Bauer suggested that it is the intervention of consciousness into the measurement process which causes the reduction.[6] And Everett's 'world-splitting' suggestion (1957) is another. According to Everett, $S + M$ being in a superposition $\Sigma \, c_i f_i \times F_i$ at t 'destabilizes' the actual world so that it splits into several alternate worlds in each of which one element $f_i \times F_i$ of the superposition survives as the state of $S + M$ at t for the various i. But since we can be aware of only one world at a given time, viz. *our* world whatever it is, this means that we effectively cause a reduction of the wave-packet: it is our ignorance of other worlds which 'causes' the reduction. Although of course in each alternate world there is a counterpart of us, aware only of their own world.

I shall not discuss these various means of accommodating the reduction of the wave-packet within QT however, because, in my view, there is no need to postulate the reduction of the wave-packet in the first place. (And here I refer back to chapter 3, section 2, where I dismissed several reasons for postulating the reduction of the wave packet.) In particular there is no need to postulate it in the present context because, as I shall now show, there is a perfectly straightforward solution to the Schrödinger cat paradox, one which simply follows from the various principles for interpreting the density operator developed in the previous chapter, and which repudiates the Copenhagen view that $Q(M)$ has an indeterminate value when $S + M$ is in a superposition of eigenstates of $Q(M)$.

From (iii) (via the von Neumann rule for the density operator of sub-systems at the end of Appendix 6) it follows that

(iv) $\hat{W}(M, t') = \Sigma |c_i|^2 \, \hat{P}(F_i).$

---

[6] Reprinted in London and Bauer (1983).

But this density operator is diagonal in the eigenvectors of $Q(M)$, and hence, according to (Bohr)' of the previous chapter, $Q(M)$ has a determinate value at t', irrespective of the fact that $S + M$ at t' is in a superposition of eigenvectors of $Q(M)$ for different eigenvalues. More particularly, taking M as Schrödinger's cat, we see that at the end of the measurement it is either determinate that the cat is dead or determinate that the cat is alive.

This last conclusion does not by itself resolve the paradox. Indeed the paradox can be seen as lying precisely in the juxtaposition of this last conclusion with the conclusion from another line of reasoning, viz. from the fact that the cat plus electron is in a superposition of cat-dead and cat-alive states to the conclusion that the cat is neither determinately dead nor determinately alive. Or, to put it another way, the paradox arises because from the derived (indeed known) fact that the cat is either determinately dead or determinately alive it seems that we can infer that the cat plus electron is in a mixture of cat-dead and cat-alive states, which *ex hypothesi* it is not, i.e. *ex hypothesi* it is in a superposition of such states. The resolution of the paradox lies in pointing out that the line of reasoning used here (in both alternatives) implicitly relies on the principle (Weaker Converse) introduced in Chapter 4., viz. that if Q has a determinate value for S at t then S at t is in a mixture of Q eigenvectors (or its contra-positive). But this (Weaker Converse) was demonstrated to fail (in Chapter 4). In short we know that the cat is either dead or alive. And it is only the invalid (Weaker Converse) which would allow us to infer from this that the cat plus electron is in a mixture of cat-dead and cat-alive states— contrary to the hypothesis that the cat plus electron is in a superposition of such states. By rejecting this (Weaker Converse) we therefore eliminate the paradox.

It is important to note that what I am suggesting here does not commit me to saying that relative to the system consisting of the cat alone the cat is either dead or alive, but that relative to the system consisting of cat plus electron it is indeterminate whether the cat is dead or alive. I explicitly avoid such relationalism by concluding that the cat is either dead or alive absolutely, i.e. not just relative to some particular system, on the basis just that there is *some* system (in this case the cat alone) the density operator for which is diagonal in cat-dead and cat-alive states. That this basis is sufficient to justify the relevant conclusion is simply a consequence of the density operator rules I have adopted which imply (Bohr)'.

It would of course be *ad hoc* to resolve the Schrödinger cat paradox in the way I have just suggested if no independent reasons were advanced for the relevant density operator rules other than that they solved the latter paradox. However, I have advanced independent reasons for adopting these rules in earlier chapters, and will give yet others. In the final analysis, however, one of my reasons for adopting these rules is that they solve various paradoxes such as Schrödinger's cat paradox. But surely my commitment to these rules is none the worse for that. If these rules do enable a solution of all the paradoxes in a better way than their rivals, e.g. by not invoking a mysterious 'reduction of the wave-packet', then surely this is a good reason for adopting them.

The resolution of the Schrödinger cat paradox just given represents a return to Bohr's original view (rather than of his Copenhagen successors) that there is no physical reduction of the wave-packet (see Jammer (1974) p. 209) and that the 'transition' to a determinate value for the measured quantity at the end of measurement is purely an artefact of a change of descriptive standpoint. In particular the views I have been suggesting deny that there is such a phenomena as the 'destruction of interference effects', as portrayed in the traditional pure-mixed reduction of the wave-packet. Such a process only appears to take place because if, at the end of the measurement of S by M, we restrict ourselves to observing physical quantities for M alone then the interference effects which are present in $S + M$ are not observable, i.e. information about them is lost in taking the trace over $\mathcal{H}(S)$; but they are still there *and* observable under appropriate probing of $S + M$.[7] (Although it should be added that Daneri, Loinger, and Prosperi, *inter alia*, do attempt to show that the destruction of interference effects is at least achieved approximately by focusing on the time-averaged nature of the macroscopic physical quantities. Cartwright, in her (1974), summarizes and identifies herself with these views.)[8]

---

[7] Cf. Jauch's somewhat similar solution of the paradoxes in his (1968). My solution differs from his because of its articulation within the axiomatic framework I have given in earlier chapters, and also because (unlike Jauch) I do conclude that the cat is either determinately dead or alive—not just that it is either dead or alive from the point of view of any measurements/observations conducted on it alone.

[8] Also Krips (1969).

4. THE CORRELATION PROBLEM

The resolution of the paradox just suggested leaves us with the following problem (posed in special form in the final section of Chapter 3). Suppose we start an immediate repeat measurement of Q in S at time t' when the first measurement of Q stops, and using a second measuring apparatus M' (with the same restrictions which we imposed on the initial measurement). Then it is easy to show from within QT that there is a perfect correlation (with probability 1) between M registering the value q_i at t' and M' registering the value q_i at t'', where t'' is the time when the repeat measurement terminates. In other words we can show that there is a perfect correlation between the results of the initial and repeat measurements.

Proof Let M' at t' be in F'_0 so that $f(S + M + M', t') = F'_0 \times \Sigma c_i f_i \times F_i$. Moreover we have that $\hat{U}_{t'', t'}(S + M + M') \, F'_0 \times f_i \times F_i = F'_i \times f_i \times F_i$ for all i if we assume that the repeat measurement is conservative of Q and not affected by the state of M at t', or vice versa, where of course F'_i is the state of M' in which it registers value q_i. As such

$$\hat{U}_{t', t}(S + M + M') \, F'_0 \times \Sigma c_i f_i \times F_i = \Sigma c_i F'_i \times f_i \times F_i.$$

But taking the trace of the projection onto this vector over $\mathscr{H}(S)$ gives

$$\hat{W}(M + M', t'') = \Sigma |c_i|^2 \, \hat{P}(F'_i \times F''_i)$$

(by using (von Neumann) of Appendix 6). And this means that at t'' there is probability $|c_i|^2$ that both $Q(M)$ has its ith value *and* $Q(M')$ has its ith value, for all i (by (Bohr)'). Hence, since $\Sigma |c_i|^2 = 1$, any event other than the joint event of $Q(M)$ and $Q(M')$ together taking their i^{th} values, for some i, has zero probability at t''. In particular there is zero probability at t'' of $Q(M)$ taking its ith value and $Q(M')$ taking its jth value, for $j \neq i$. Hence the stated correlation obtains. QED.

How is the latter correlation to be explained? It is of course easy to explain it if we assume the reduction of the wave-packet; i.e. if

$$\hat{W}(S + M, t') = \Sigma |c_i|^2 \, \hat{P}(f_i) \times \hat{P}(F_i)$$

(as the pure-mixed reduction of the wave-packet requires) then, by the given density operator rules, there is probability $|c_i|^2$ of the joint event that Q has value q_i and $Q(M)$ has its ith value at t', for each i.[9] And hence, since $\Sigma |c_i|^2 = 1$, there is zero probability of any other

---

[9] For the pure-pure reduction the correlation can also be explained.

combination of Q and $Q(M)$ values at t'. In other words at t' not only does $Q(M)$ have its ith value (i.e. registers values q_i) for some i, but also Q has some value, *and* there is a perfect correlation (with probability 1) between Q having q_i in S at t' and $Q(M)$ having its ith value at t', for all i. We then explain the correlation between the initial and repeated measured values in terms of this earlier correlation at t' together with an assumption of 'passivity of measurement' (that the registered value reflects the value possessed by the measured physical quantity if there is one to reflect).

Without the reduction of the wave-packet this explanation of the correlation between the initial and repeat measured values seems to be unavailable. But this is not the case. In order to explain the target correlation between the repeat and initial measured values we do not need to postulate anything as strong as the reduction of the wave-packet. All we need is the weaker assumption that there is a correlation between the values possessed by Q at t' and the measured values registered by M at t'. And we can indeed derive just such an assumption from the form for $\hat{W}(S + M, t')$, albeit by a 'circular' route which simply reverses the logical order of the explanation (by arguing back from *explanandum* to *explanans*) as follows. From the *explanandum* that there is a perfect correlation between the possible value for Q which is registered by M at t' and the result of measuring Q in S at t' (not t), and assuming passivity of measurement, it follows that there must be a perfect correlation between the possible value for Q which is registered by M at t' and the value possessed by Q in S at t'.[10] And the latter correlation can then be used as an *explanans* to explain the target correlation (between the initial and repeat measured values) with the help of passivity of measurement. To point out the circularity here—of *explanandum* justifying *explanans*—does not detract from the explanation. Such circularity is perfectly permissible in the context of explanation albeit not in the context of justification. (For example consider Scriven's famous example of explaining the collapsing bridge by appeal to metal fatigue, where the only evidence for the metal fatigue is the very fact to be explained, viz. the collapse of the bridge.)[11] Indeed the only difficulties for this explanation arise

---

[10] The derivation here is not as straightforward as might seem at first sight. In particular the passivity condition required is statistical, asserting that the conditional probability of the measured value of Q being r given that a measurement of Q takes place when Q has the value r is 1. For details of this see Krips (1987).

[11] See Scriven (1965).

from criticisms of the model for measurements which I have been using here (see next section), and from doubts which can be cast on the principle of passivity of measurement (see Chapter 7).

5. DERIVING THE BORN INTERPRETATION

So far in this chapter I have defined 'ideal measurement' in the context of QT, and resolved the generalized Schrödinger cat paradox for the measurements so defined without appeal to a process of reduction of the wave-packet. Indeed I have repudiated the existence of such a reduction process in favour of Bohr's view that the transition to a classical description for the measuring apparatus is a mere artefact of a change of mode of description.

I now point out that (iv) (viz. that $\hat{W}(M, t') = \Sigma |c_i|^2 \hat{P}(F_i)$), which was derived above for a particular model of the measuring process, straightforwardly enables the derivation of (BSI) for that particular model measurement process if we avail ourselves of the various density operator rules which I have proposed. From (iv) and (Bohr)' it follows that there is probability $|c_i|^2$ of $Q(M)$ having its ith value at t', and hence, by (Cal)', there is probability $|c_i|^2$ of M registering the value q_i at t'. Thus we see that for a conservative measurement of a non-degenerate physical quantity Q for S at t there is probability $|c_i|^2$ of measuring Q to have the value q_i if $f(S, t) = \Sigma c_i f_i$ (where f_i is the eigenvector of Q for value q_i), which is just what (BSI) says.

This sort of result provides yet further justification for the density operator interpretation I developed in the previous chapter, by showing how it entails other principles (such as (BSI)) which, on more traditional axiomatic schemata for QT are taken as fundamental. Clearly the more general the class of measurements for which we can derive (BSI) in this way, the stronger is this justification; and in Appendix 9 I shall show how one such generalization is possible.

I note also that a more general derivation of (BSI) is needed because Araki and Yanase (1960) have shown (following a proof of Wigner (1952)) that in certain quite common cases the basic condition:

$$\hat{U}_{t', t}(S + M) F_0 \times f_i = F_i \times f_i' \quad \text{for all } i,$$

which I took to be characteristic of ideal measurements, can be satisfied only approximately (even for non-conservative measurements with $f_i' \neq f_i$). This means that the derivation of (BSI) must be implemented for more general models of measurements than those considered so far if it is to have any significance (and in particular if the class of ideal measurements is not to be vacuous). Moreover it is necessary to develop such models if the viability of QMT (quantum theory of measurement) and hence arguably of QT itself, is to be sustained. After all if QT implies that measurements of q-quantities are not possible then it is QT itself which is under threat.

In the next section of this chapter I shall address another of the traditional paradoxes of QT, the Einstein–Podolski–Rosen paradox; and show how it too is resolved by the interpretation being suggested here.

6. EINSTEIN–PODOLSKI–ROSEN PARADOX

In their famous (1935) paper EPR argued for the 'incompleteness' of QT. The interest of that paper lies not so much in the argument itself, which I shall show to be unsound. Rather the interest lies in the argument's challenge to various Copenhagen doctrines, and in its support for those interpretations of QT which allow incommensurable q-quantities to have determinate values simultaneously (eg 'hidden variable' interpretations of QT as well as the interpretation of QT I put forward in the previous chapter).

EPR consider a pair of spatially well-separated electrons S_1 and S_2 prepared in a particular state F at t which is not only an eigenstate for value 0 of $P_1 + P_2$ (the sum of the momenta of S_1 and S_2 along some particular direction), but is also an eigenstate for value 0 of $Q_1 - Q_2$ (the difference between the position coordinates of S_1 and S_2 along that same direction). This state F is possible because $P_1 + P_2$ commutes with $Q_1 - Q_2$, and hence they share a complete set of eigenvectors, despite P_1 and Q_1 individually not commuting, and similarly for P_2 and Q_2. It is taken that none of the P_1, Q_1, P_2, or Q_2 is actually measured at t. (Note that it is only a convenient fiction here to talk of eigenstates for $P_1 + P_2$ or $Q_1 - Q_2$, since they have continuous spectra; but I follow EPR in this fiction. It is a harmless fiction in the present context, because the whole argument can be reformulated in terms of spin components which do have proper eigenvectors—see Bohm and Aharanov (1960).)

Now EPR point out that since F is an eigenstate of $P_1 + P_2$ we can deduce from QT (from (BSI) in particular) that, for any p, 1 is the value taken by the conditional probability of P_1 being measured to have value $-p$ for S_1 at t conditional on P_2 being measured to have value $+p$ for S_2 at t (and also of course conditional on P_1 being measured for S_1 at t). This anti-correlation between measured values for P_1 and P_2 at t means that there is a possibility of an *indirect* measurement of P_1 at t which does not disturb S_1, viz. by measuring P_2 at t. This indirect measurement of P_2 at t does not disturb S_1 at t because *ex hypothesi* S_1 and S_2 are well separated at t, and because we make the assumption (A) there is no instantaneous action at a distance.

EPR then make the first controversial step in their argument. They argue that the mere possibility of measuring P_1 at t by this indirect method of measurement means that there must be some 'element of reality corresponding to P_1 at t' (as they put it).

We can defend this first step in their argument by assuming that EPR are taking as a background assumption that (B) measurement of P_1 in S_1 at t is only possible by *either* 'creating' (in some sense) a value for P_1 in S_1 at t which the measurement then reports *or* by passively reporting an 'independently existing' value for P_1 in S_1 at t ('independently existing' meaning 'existing whether or not the measurement were to take place'). But (as we argued above) the measurement of P_1 at t (being indirect) cannot disturb S_1 at t, i.e. can have no effect on S_1 at t, and hence the first of these possibilities is ruled out. Therefore the second possibility obtains, i.e. P_1 at t must have a value independently of whether the measurement takes place; and in that sense the value of P_1 in S_1 at t is 'objective', or 'corresponds to an element of reality', as EPR put it. Note that there is an extra hidden assumption at work here in ruling out the first of these alternatives, viz. the assumption that (B)' if the act of measuring P_2 on S_2 at t creates (brings into existence) the value possessed by P_1 at t then there must be a *causal* relation between measuring P_2 on S_2 at t and P_1 possessing a value at t.

The above assumption (B) does of course simply beg the question against Bohr's interpretation of QT, because (I claim) Bohr allows a measurement to take place which reports no possessed value at all, be it created by the measurement or not. In other words I claim that for Bohr the measured value may emerge without reflecting a possessed value whether that possessed value exists independently of the measurement or is created by it. Note however that this last claim does go against Feyerabend's interpretation of Bohr in Feyerabend's

(1962). That is according to Feyerabend, Bohr sees the characteristics of measured systems as 'brought into existence' by the measurement, albeit not in a causal way but rather in the way that my being taller than average may be brought into existence by everyone else becoming shorter, i.e. Feyerabend sees Bohr as advocating a relational view of the characteristics of q-systems. On that view, Bohr would have to agree with EPR that measurement does report a possessed value, although, by contrast with EPR, he would see that possessed value as 'created' by the fact of measurement. The creation here would not be causal however, and hence not constitute instantaneous action at a distance. Thus, on Feyerabend's view, Bohr's disagreement with EPR would be over (B)′, whereas I see Bohr's disagreement with EPR being over (B). Either way however the first step in the EPR argument is rejected by Bohr. Nevertheless we shall go along with EPR in their first step, because on the interpretation of QT which I suggested in the last chapter this first step can be made good (albeit not by appeal to the dubious background assumptions (B), (B)′). It can be made good because $\hat{W}(S_1, t)$ is the identity operator on S_1; and, according to (Bohr)′, this means that *every* q-quantity for S_1 has a determinate value at t, and, in that sense at least, corresponds to an element of reality just as EPR claim.

The second controversial step in the EPR argument is the putting forward of a 'completeness condition'. EPR claim that for QT to give a complete (*sic*) representation of reality requires (as a necessary condition only) that whatever corresponds to an element of reality—such as the value of P_1 at t—must be appropriately represented within QT, and in particular that P_1 at t must be represented within QT as having a 'definite' value. (What is meant by 'definite' here will be discussed below.) And similarly it is argued that if QT is to be complete then QT must also represent Q_1 at t as having a definite value, because there is also a possibility of indirectly measuring Q_1 at t which arises because F is also an eigenstate of $Q_1 - Q_2$. Hence P_1 and Q_1 must *both* be assigned definite values at t by QT if QT is to be complete. The third and final step in the argument points out that the uncertainty principle, which is part of QT, precludes both P_1 and Q_1 having definite values; and so QT, it follows, is not complete.

Note that this argument does not rely on the *compossibility* of indirectly measuring P_1 and Q_1, but only on there being both a possibility of indirectly measuring P_1 at t and a possibility of indirectly measuring Q_1 at t.[12] Indeed if Bohr is right then the indirect measurements of both P_1 and Q_1 at t (*qua* the joint direct measurement of P_2 and Q_2 at t) is not possible.

There is, however, a fallacy in either the second or third steps in this argument. We can expose this fallacy by examining what EPR could possibly mean by the term 'definite value' as it appears in their argument. If 'P_1 has the value p *definitely* in S_1 at t' is given a strong interpretation as 'P_1 has the value p with *certainty* in S at t' then indeed the third step in the EPR argument is valid. (Because it is indeed the case that P_1 and Q_1 cannot both have particular values with certainty in S_1 at t, this being disallowed by the uncertainty principle, and more generally by the fact that $f(S_1, t)$ cannot both be an eigenvector of P_1 and Q_1.)[13]

But under this strong construal of 'definite' it is not clear that EPR's completeness condition is plausible. After all why should P_1 at t be represented within QT as having a particular value *with certainty* just because it corresponds to an element of reality? More generally why should the mere possibility of being able to predict an event with certainty (without disturbing it) imply that the event really is certain, i.e. has probability 1? After all it may well be possible to predict with certainty what side a penny will come down on were we to know enough about how it is tossed on some occasion, but this does not mean that the probability of its coming down heads, say, is other than one-half. And by parity of reasoning there is no reason why the possibility of knowing with certainty what value P_1 has at t implies that it must have that value with certainty. Surely at most it implies that P_1 at t has some particular value or other, with perhaps a statistical distribution over the various possible values, but without any of the values necessarily having probability 1? In short EPR need to give us more argument here if they are to support their completeness condition under a strong construal of 'definite'.

On the other hand 'P_1 has value p definitely in S_1 at t' may be merely taken to mean 'P_1 has some value or other in S_1 at t'. In that case, as just indicated, EPR's completeness condition becomes more plausible (although it is still by no means crystal clear!). But then the

---

[12] The compossibility of two propositions is clearly stronger than the conjunction of the possibility of each: thus p and $-p$ are not compossible, but each may well be possible.

[13] Since P_1 and Q_1 are strongly incompatible, i.e. share no eigenvectors.

third step in the EPR argument is no longer cogent. In particular the step which claims that the uncertainty principle does not allow both P_1 and Q_1 to have 'definite' (now meaning particular) values becomes questionable (although this is not to deny that QT precludes P_1 and Q_1 from both having particular values *with certainty*).[14]

Although interestingly enough under the Bohr interpretation, there is a justification for this third step in the EPR argument (since, for Bohr, for Q to have some value at all, meaning some *determinate* value, implies having it with certainty). But as indicated above it is already the first step of the EPR argument which Bohrians find question-begging.[15] So it seems that whatever EPR might mean by 'definite', and whatever interpretation of QT is followed, the EPR argument to the effect that QT is incomplete fails. This theme of the incompleteness of QT will however re-emerge in less tendentious form when we consider the Kochen and Specker theorem later.

Finally on the topic of EPR argument, I point out that the situation it explores brings out some significant features of QT, and of QMT in particular. First of all it makes clear that, *contra* one Copenhagen doctrine, simultaneous measurements of momentum and position (in the same direction on the same particle) are compossible, i.e. we can both *indirectly* measure P_1 (by measuring P_2 directly) and *directly* measure Q_1 at the same time t. And this in turn can be seen to imply the invalidity of the pure-pure reduction of the wave-packet formula when it is interpreted as a general principle characteristic of all measurements (direct or indirect). That is if we assume that the reduction of the wave-packet formula applies to all measurements then the simultaneous measurements of P_1 and Q_1 at t as portrayed in EPR imply that at the end of those measurements the state of S_1 is both an eigenvector of P_1 and of Q_1. But P_1 and Q_1 share no eigenvectors, and hence, by *reductio*, we see that the reduction of the wave-packet formula must fail for at least one of those measurements.

However there is a way to avoid this conclusion. In particular the premiss (first controversial step in the EPR argument) that both P_1 and Q_1 are elements of reality can be questioned. This can be questioned if we allow instantaneous action at a distance so that the measurement of P_2 at t may disturb S_1 at t and hence we no longer have a guarantee that the measured value of P_1 at t can be predicted

---

[14] On this point see the discussion in Jammer (1974), ch. 3, especially p. 83.
[15] See Bohr's response to EPR (1935).

with certainty (viz. by measuring P_2 at t) but *without disturbing S_1*. (Jauch can be seen as taking this sort of a line, by questioning the isolation of S_1 and S_2 (1968) section 11.10.) By taking this view we can reconcile QT's completeness with Einstein's definitions of completeness and physical reality.

It is interesting that it is non-local action which in this last paragraph performs the role of rescuing the completeness of QT in the face of Einstein's definitions, because as we shall see it is precisely such non-local action which, it seems, is indicated by realist interpretations of QT in the light of Bell's theorem. Indeed this point is more than interesting: it would seem to tell strongly against such 'hidden variables' interpretations of QT. It tells against them because one rationale for such 'hidden variables' interpretations is precisely the EPR argument in its role as demonstrator of the incompleteness of QT (an incompleteness which is 'completed' by the hidden variables). But if we allow in non-local action then this EPR demonstration of incompleteness fails from the outset (see assumption (A) introduced at the beginning of the EPR argument), and so the rationale for the introduction of hidden variables is removed. Thus non-local hidden variable theories would seem to be self-defeating, undercutting one of the central rationales for adopting a hidden variables theory in the first place. And it is, it seems, just such non-local hidden variable theories which we are forced to adopt if we take hidden variables seriously. These questions will be addressed further in the final two chapters.

7. THE PROJECTION POSTULATE

For those who have found my interpretation of QT here too radical in its rejection of the reduction of the wave-packet let me finally note that there is at least one consequence of the pure-mixed reduction of the wave-packet formula which is preserved under my interpretation of QT. It is the principle (which I shall call 'the projection postulate'):

(Proj)   Measurement of Q in S at t projects S into a mixed state, viz. a mixture of eigenvectors of Q.

This principle, unlike the reduction of the wave-packet, is left untouched by the fact that simultaneous measurements of P_1 and Q_1

may take place when $S_1 + S_2$ is in the EPR state at t. This is because taking the trace in $\mathscr{H}(S_2)$ of $\hat{W}(S_1 + S_2, t)$ (i.e. of $\hat{P}(F)$) shows that $\hat{W}(S_1, t)$ is the identity operator $\hat{I}_1$ on $\mathscr{H}(S_1)$, and this is both a mixture of momentum eigenvectors and of position eigenvectors, (i.e. $\hat{I}_1 = \Sigma \hat{P}(f_i)/N_1$ for any basis set $\{f_i\}$ in $\mathscr{H}(S_1)$, where N_1 is the dimension of $\mathscr{H}(S_1)$.)

This 'projection postulate' must however be distinguished from what I above called 'the reduction of the wave-packet'. The latter imposes a condition not just on the state of the measured system (as the projection postulate does), but on the state of the combined measuring apparatus and measuring system. For example the pure-mixed reduction of the wave-packet has it that

$$\hat{W}(S + M, t') = \Sigma |c_i|^2 \hat{P}(f_i) \times \hat{P}(F_i),$$

whereas what I called 'the projection postulate' has it merely that

$$\hat{W}(S, t') = \Sigma |c_i|^2 \hat{P}(f_i).$$

It is only the former I reject, although I agree that the latter is only valid for a small (but historically rather central) class of measurements—the conservative ones (which, as we have seen at the end of section 3, are unfortunately the very class of measurements upon which the Wigner–Araki–Yanase proof casts doubt).

Note again that my view of what I am here calling 'the projection postulate' is quite Bohrian in spirit in that I do not take it as describing a physical process separate from the Schrödinger interaction between measuring apparatus and measured system.[16] The 'projection' of the measured system S into a mixture by the measurement process is merely an artefact of a change of standpoint from considering $S + M$ to considering S alone at the end of the measurement interaction.

[16] See the discussion in Jammer (1974) p. 198.

6

Realism

Determinacy; passivity; no-disturbance; ideal measurements.

ALREADY in this book one substantial issue has been discussed under the rubric of 'realism' in connection with QT, viz. the issue of whether the state-vectors and density operators of QT represent 'real' properties of q-systems. For the purposes of this chapter I take this particular issue to have been settled in favour of realism. But more generally what does it mean to adopt a 'realist interpretation' of QT? I shall take it to mean accepting that QT is true, that the objects QT refers to (electrons, protons, etc.) exist, that the properties it refers to are 'real', and in particular that the physical quantities it refers to are 'real'; in short it also means that we can interpret QT 'literally' in the sense that we can take all its referential terms as genuinely referring and not just as convenient fictions or metaphors for the real.

A 'realist interpretation' in this latter sense is rather more robust than the sort of interpretation we are in fact willing to extend to many of the theories we accept. For example I may wish to accept the account of space-time structure put forward by general relativity, expressed in terms of a metric function defined over the domain of space-time points. Moreover I may be quite realist about this structure, *qua* structure, taking it to describe real properties of the space-time manifold or at least of the objects located in that manifold (albeit with some degree of arbitrariness). Nevertheless I may wish to repudiate the existence of space-time points, to take them as mere conveniences for representing these structural properties.[1] Such a view involves only a partially realistic interpretation of general

---

[1] See van Frassen (1980), ch. 3. for example.

relativity in the sense that I am using the term 'realist' here. In terms of this distinction between partial and total realism we see that what I have been advocating here is a realist (not just partially realist) interpretation of QT. By contrast Cartwright (1982) and Bohr advocate a partially realist interpretation of QT (realist about electrons but not about their states as represented by state-vectors).

There is however another issue which can be classified under the heading of 'realism' in connection with QT. Consider the realist metaphysical view that there is an independently existing objective reality which we aim to describe in science. QT, or even an interpretation of QT, cannot of course prove this metaphysics. But a particular interpretation of QT can support this realist metaphysics by presenting QT as descriptive of an objective reality. In particular the metaphysical realist cause gains support from the fact that QT is not interpreted instrumentally under the interpretation I have been arguing for, *and* that the reality described within QT is 'objective' in the following two respects. First, physical quantities have values independently of whether they are measured, let alone observed, to do so. (This is so even for some versions of the 'orthodox interpretation' which allow values to be assigned to physical quantities just as long as the relevant systems are in eigenstates of the relevant quantities.) Moreover the theoretical entities of QT, e.g. the state-vector, the probabilities, and the indeterminacies, are taken to be 'objective' in the sense of not being (in a logical sense) mind-dependent. For example the probabilities are not taken as epistemic, and *a fortiori* the wave-packet is not seen as collapsing on the basis merely of a change in our knowledge.

The interpretation of QT I have been arguing for here can be seen as supporting realism in yet a third way, viz. by repudiating the transcendental idealism implicit in Bohr's views of the relation between science (indeed knowledge in general) and the world. That is for Bohr the categories in terms of which QT is to describe the world—the 'phenomenal world'—are predetermined by the categories of the understanding (and ultimately by the categories of perception) rather than developing out of the world as it is. The realist interpretation I have been arguing for here (following Heisenberg, Born, *et al.*) repudiates this idealism.

But traditionally there has been more to a realist interpretation of a theory than interpreting it 'literally' in the sense just given, or even taking it as an instance of an anti-subjectivist anti-idealist metaphysical programme. In particular in the case of QT it is traditional[2] to take the endorsement of certain principles as characteristic of a realist interpretation of QT. They are:

(Det Q)   No q-quantities ever have indeterminate values.

(Pass Q)   For ideal measurements the measured value of a q-quantity (i.e. the value reported by the measurement) = the possessed value of that quantity just prior to the measurement.

(NDQ)   The possessed value of a q-quantity just after measurement = the measured value, for ideal measurements.

(The ideal measurements referred to here are those stipulated within QMT.)

Why should these three principles be taken as necessary for a realist (*sic*) interpretation of QT? I shall now examine some of the answers to this question which might be given/have been given in the literature. I shall end up rejecting all of the answers, and with them the view that (Det Q), (Pass Q), and (NDQ) have anything much to do with realism (*sic*) in the context of QT.

Nevertheless I shall discuss the validity of all these principles in the next chapter. After all, since these three principles are extensions of certain central principles of classical metaphysics to QT, the question of their validity must be confronted as part of the more general question of whether a classical metaphysics can be grafted onto QT. I shall also continue to refer to (Det Q) and (Pass Q) as 'realist principles' in later chapters, if only to conform with the literature.

First of all I shall consider what seems to be Popper's argument for connecting (Det Q) with the issue of realism. If (Det Q) fails then QT refers to indeterminacies. But (for Popper) indeterminacies are 'subjective'. Hence if (Det Q) fails then QT, literally construed, is seen to be asserting the presence of subjective elements in the domain of science. And that contradicts the metaphysical doctrine of objectivism,[3] viz. that science concerns itself with objective states of affairs, and hence can be seen as contradicting realism in its guise as objectivism. (Note that a similar argument might be mounted on the basis of the presence of probabilities in QT, if one took the probabilities as epistemic and hence subjective.)

---

[2] See Healey (1979) for example.   [3] See Popper (1982c), vol. 3, pp. 144ff.

But Popper's argument here fails because its premiss that indeterminacies are subjective can be questioned. It is perhaps true that certain proponents of the orthodox interpretation of QT, notably the early Heisenberg of (1930), did make out indeterminacies to be subjective in the relevant sense: they took them to be aspects of the state of knowledge of observers (so that to say that the value of Q in S at t is indeterminate to a certain extent is to say how accurately we know the value of Q in S at t). And, under that construal, it is indeed the case that if QT refers to indeterminacies then objectivism has failed: after all, for an objectivist, physics is not about the state of knowledge of observers but about the objects of that knowledge.

Not all proponents of the orthodox interpretation of QT have followed Heisenberg's (1930) view however. Indeed the whole point of the distinction which appeared in the literature between '*unsicherheit*' ('uncertainty') and '*unbestimmtheit*' ('indeterminacy' or 'indefinability') was to disavow precisely this view (see Jammer (1974) p. 61). Thus Popper's use of 'subjective' here does not apply to all authors in the history of QT. It was only really the followers of Heisenberg (1930) who took indeterminacies as 'subjective' in a sense which would license an inference from admitting indeterminacies into QT to the failure of objectivism.

Of course to say this is not to imply that it is a slight matter for indeterminacies to be construed realistically within QT. Far from it. Classical metaphysical views to the contrary are deeply embedded within our web of belief. What I am pointing out however is that there is no reason to tie realism to a ban on reference to indeterminacies.

It has also been traditional to take (Pass Q) as essential to a realist interpretation of QT.[4] More generally, some sort of principle of measurement as passive has often been taken as constitutive of the notion of an ideal measurement, both within and outside the context of QT:

(Pass)   For ideal measurements the possessed value of the physical quantity just prior to its measurement = the measured value, i.e. ideal measurements are 'faithful'.

I shall argue against both of these claims. Not only do I wish to deny that (Pass Q), as a special instance of (Pass), is necessary to a realist

---

[4] See Healey's 'realist' principle (FM) in (1979).

interpretation of QT, but also to deny that (Pass) is true by definition of 'ideal measurement'. I consider the second of these issues first of all.

The first point to make (in arguing against the definitional status of (Pass)) is that there are at least two principles, other than (Pass), which have claims to be definitive of the notion of an ideal measurement, viz.

(ND)   For ideal measurements, the possessed value of the measured physical quantity just after measurement = measured value.

and:

(Pass)′   A physical quantity has a particular value with probability 1 iff any ideal measurement reports that value with probability 1, i.e. ideal measurements are statistically faithful at least in the extreme case of probability 1.

(Prima facie this last principle may seem to be a mere weakening of (Pass), but it is only if we interpret the probabilities as actual relative frequencies that this is so.)

In classical metaphysics of course, all of these various principles (Pass), (Pass)′, (Det), and (ND) are true of ideal measurements together; and so the issue of which if any is 'really' definitive of ideal measurements does not arise in practice. But let us now press that issue. In particular consider a domain of theoretical discourse within which it is guaranteed that the members of some independently characterized class of measurement processes C are statistically faithful in the sense of (Pass)′. ('Independently characterized' here means characterized in terms other than that of being statistically faithful.) Such a guarantee can be obtained by deriving from the accepted laws in that domain that all measurements of the form in question are statistically faithful. But suppose also that there is no independently characterized class of measurements within the domain in question which we can guarantee to be faithful *simpliciter* (in the sense of (Pass)). This does not mean that all, or even some, of these other measurements can be shown to be unfaithful. All it means is that we cannot give an independent characterization, within the resources of the relevant domain of discourse, of a class of measurements which are faithful *simpliciter*. In such a case we could perhaps persevere with taking (Pass) as characteristic of ideal measurements, by insisting for example that only some of the members of C are ideal, viz. precisely those which satisfy not only (Pass)′ (which all the members of

C do, *ex hypothesi*) but also (Pass), although we would of course have to admit ignorance of which measurements in *C* were ideal.

Suppose however that it was also derivable from the accepted laws for the relevant domain of discourse that no measurement in *C* could satisfy (Pass). Would we then say that there was no class of ideal measurements for this domain? In such a case, it would seem reasonable *faute de mieux* to take the members of *C* (given that they are at least statistically faithful) to be ideal—ideal in the extended sense that they are the measurements for which there is the best available guarantee of reliability. The members of some other class of measurements (other than *C*) may well turn out to be more faithful on occasions, indeed even often, but we have no guarantee of this superior reliability, i.e. no general recipe to pick out those measurements which are more reliable; and as such their hypothetical superior reliability is of no practical use. It in no useful way makes them 'better'.

Accepting all of this we see that a new notion of ideal measurement may become appropriate in new theoretical domains, with (Pass)′ as definitive, and this opens the possibility of (Pass) failing by (Pass)′ providing an independent purchase on the notion of ideal measurement. Indeed these considerations suggest that even in the first case we considered, for which an unknown number of members of *C* satisfied (Pass) and for which, more generally, there was no independently characterized set of measurements known to satisfy (Pass), we should take the members of *C* as our ideal measurements. In particular if we can prove theoretically within some such domain of discourse that at least some possible instances of ideal measurements, as defined by (Pass)′, are unfaithful then (Pass) will be said to fail. This will be so even if we do not know for which of the measurements that satisfy (Pass)′ it is the case that (Pass) fails. (Indeed it is arguable that such ignorance is essential. If we knew which measurements were faithful then we would presumably separate them off as *the* ideal measurements.)

This last example is not just of academic interest. Within QT there is no generally characterized class of measurement processes which can be guaranteed to be faithful; and generally (Pass Q)′ can be seen as taking over the role of defining ideal measurements. (It is taken to do so within most texts on foundations of QT—see, for example, d'Espagnat (1971) Chapter 11). Note however that on other accounts (NDQ), or a cognate principle, the 'projection postulate', takes over this role;[5] but as we have seen in Chapter 5 there are difficulties for either of these suggestions.)[6] Moreover, as we shall see in the next chapter, the Kochen and Specker theorem can be used to show that (Pass Q) must be rejected for the ideal measurements defined by (Pass Q)' if (Det Q) is also accepted. And if (Det Q) is not accepted then (Pass Q) must still be rejected, albeit for different reasons.[7] Either way therefore (Pass Q) must be rejected. In other words it seems that QT prevents all possible ideal measurements from being faithful *simpliciter* if the ideal measurements are taken to be statistically faithful, which is just the possibility we were entertaining in the previous paragraphs.

But however we decide the status of (Pass Q)—whether or not it is taken as definitive of ideal measurements—there remains the question of whether it is essential to a realist interpretation of QT; and it is to that question which we now finally turn. It is a traditional part of realist concepts of science that within science we ought to develop a 'mirror for nature' (see Rorty (1980) for example): that scientific theories ought to mirror what goes on in the real world. If we take the existence of such a mirror to consist of at least the possibility of ideal measurements for all physical quantities then we see that the satisfaction of (Pass Q) can be taken as a prerequisite for QT to satisfy realist precepts, i.e. to sustain a realist interpretation. So it seems that (Pass Q) can legitimately be identified as essential to a realist interpretation of QT (even if it is not constitutive of the notion of an ideal measurement).

But the latter argument relies on taking too literally the phrase 'a mirror for nature'. The 'mirror' here is only a metaphor for the accessibility of the 'external' world to scientific investigation; it is not more narrowly a metaphor for ideal measuring apparatuses. Once we realize this we see that the failure of (Pass Q) is no threat to realism, as long as the failure of (Pass Q) does not block all means of access to the world, which it clearly does not if (Pass Q)' is satisfied.

---

[5] e.g. von Neumann (1955), ch. 6.

[6] A class of measurements were defined which were not conservative and hence did not satisfy (NDQ). Indeed the Wigner–Araki–Yanase result suggests that the requirement of being conservative, and hence satisfying (NDQ), is too strong to be consistent with QT.

[7] Essentially because measured values are always determinate, and hence in giving up (Det Q), i.e. in allowing indeterminate values for measured quantities, we give up the possibility of always matching measured with possessed values.

Thus we see that either (or both) of (Pass Q) and (Det Q) may be dropped (and the same holds for (NDQ))[8] without prejudicing a realist interpretation of QT. In the following chapters I shall discuss the arguments for dropping these principles, and *a fortiori* for the failure to graft QT onto the framework of classical metaphysics. As indicated above however, I shall continue to refer to (Pass Q) and (Det Q) as 'realist principles', but (as we have seen) this is for convenience rather than out of any conviction that they have much to do with realism as such. In short, our interest in whether the 'realist' principles (Pass Q) and (Det Q) are consistent with QT arises not so much from an interest in whether QT is consistent with realism, but rather from an interest in the extent to which QT violates classical metaphysics.

[8] We saw in Chapter 5 that (NDQ) fails too.

7

Kochen and Specker

Gleason's theorem; Kochen and Specker; incompleteness and de-Ockhamization; the failure of passivity; contextualization of measured values; indeterminism; Redhead and Hellman; intrinsic probabilities; varieties of passivity.

1. DERIVING INCOMPLETENESS (A)

In this chapter I shall discuss some difficulties for a realist interpretation of QT. In particular I shall discuss the difficulties in reconciling the two classical metaphysical principles (Det Q) (that for any q-system S all the q-quantities for S have unique determinate values at all times t during the lifetime of S) and (Pass Q) (that the ideal measurements of QT are passive) with various of the uncontroversial laws of QT. These difficulties have been brought to focus thanks to two powerful theorems in the mathematical theory of Hilbert spaces proved by Gleason (1957) and by Kochen and Specker (1967).

Gleason's theorem has the following interesting corollary. Consider the points on a sphere. It is impossible to associate either a 1 or a 0 with each of those points in such a way that 1 is the sum of the values associated with any three points which terminate a mutually orthogonal set of radii. Formally we can state this result as follows:

(G) There is no single-valued map m from the set of points S on a Euclidian sphere onto 1 or 0 for which $m(a) + m(b) + m(c) = 1$ for any a, b, c in S which are mutually orthogonal.[1]

This result is easily generalized to spheres of higher dimension (although it does not hold for dimension 2) as well as to spheres in a

[1] Belinfante discusses various proofs of (G) which can be found in the literature (1973). Bell's is the simplest in a special form which makes it relevant to QT (1966).

complex Euclidean space, and hence to a complex Hilbert space. In generalized form it says:

(G)′   There is no single-valued map m from the set of normalized vectors in an at least three-dimensional Hilbert space onto 1 or 0, for which $\Sigma_i m(f_i) = 1$ for any complete mutually orthogonal set of normalized vectors $\{f_i\}$.

I shall now use this result as part of a *reductio* argument to show that if we assume (Det Q) then we must give up the completeness of QT. Suppose that (Det Q) holds. Then for any Q, S, t, there is a unique number q for which Q in S at t has the value q—call this unique number '$Q(S, t)$'. (I am here implicitly taking 'Q', and later 'Q'', to denote q-quantities for S, and letting 't' refer to a time t during the lifetime of S.) Note that it is the further assumption, implicit in (Det Q), that Q in S at t has a unique value which is *determinate*. This allows us to use or-elimination as an inference rule from 'Q has value q_1 or value q_2 or ... ' as premiss.[2] In other words it is this determinacy of value which allows us to argue by cases: to prove some proposition p by showing that p holds for each possible value of Q separately. This method of proof figures both in the proofs of this chapter and the next. I shall simply take this for granted from now on, i.e. that $Q(S, t)$ is not just the unique value possessed by Q in S at t but is a determinate value in the relevant sense.

Given (Det Q) we can construct a map m with range consisting of just 1 and 0 and domain within the Hilbert space $\mathcal{H}(S)$ for S, as follows. If there is some non-degenerate Q for which f is the unique eigenvector of Q for eigenvalue q and $Q(S, t) = q$ then let $m(f) = 1$; and if there is some non-degenerate Q' for which f is the unique eigenvector of Q' for eigenvalue q', where $Q'(S, t) \neq q'$, then let $m(f) = 0$.

As so far specified, however, the map m may not be single-valued because a particular f may be the eigenvector of two different physical quantities Q and Q' for eigenvalues q and q' respectively say, and we may have $Q(S, t) = q$ but $Q'(S, t) \neq q'$. In that case $m(f)$ is both 1 and 0. We eliminate this possibility by assuming:

(U)   For any S, t and any non-degenerate Q and Q', if Q and Q' have the same eigenvector in $\mathcal{H}(S)$ for eigenvalues q and q' respectively then $Q(S, t) = q$ iff $Q'(S, t) = q'$.

---

[2] As discussed in Chapter 1, p. 33.

If in addition to (Det Q) and (U) we also assume:

(C)   For any complete orthonormal set of vectors $\{f_i\}$ in $\mathcal{H}(S)$ there is a non-degenerate q-quantity for S which has $\{f_i\}$ as its eigenvectors,

then it also follows that m has the property:

$\Sigma_i m(f_i) = 1$ for any complete orthonormal set of vectors $\{f_i\}$ in $\mathcal{H}(S)$, and $m(f_i) = 1$ or 0 for all i.

Proof   Let $\{f_i\}$ be any such set of vectors. Then there is a non-degenerate Q with $\{f_i\}$ as its eigenvectors (by (C)). But, from (Det Q), $Q(S, t) = q$ for some number q, and q must be one of the eigenvalues of Q—let it be q_i for eigenvector f_i.[3] Hence $m(f_i) = 1$. But since Q is non-degenerate, i.e. has no repeat eigenvalues, and since $Q(S, t) = q_i$, it follows that $Q(S, t) \neq q_j$ for any $j \neq i$. Hence $m(f_j) = 0$ for any $j \neq i$. Hence $\Sigma_i m(f_i) = 1$ and $m(f_i) = 1$ or 0 for all i. QED.

(Note that this previous proof is a proof by cases. We prove $\Sigma m(f_i) = 1$ by showing that this holds for each possible assignment of values to Q in S at t, i.e. for $Q(S, t) = q_i$ for any i. As indicated above it is the determinacy of the value of Q in S at t which is being used implicitly here.) In short given (U), and hence that m is single-valued, m has precisely the properties which (G)′ says it cannot have, at least not for those S for which the dimension of $\mathcal{H}(S) \geq 3$. And since according to QT there are such systems S (e.g. electrons) for which dimension $\mathcal{H}(S) \geq 3$, it follows by *reductio* that we must surrender (U) or (C) or the initial realist assumption (Det Q) if we are to preserve QT intact.

At first sight (C) seems the obvious candidate for rejection by a realist who wishes to retain (Det Q) and QT; indeed in earlier chapters we have already indicated that (C) is controversial. But the second theorem we mentioned above, viz. the Kochen and Specker (KS) theorem, shows that the non-existence of m as described in (G)′ can be demonstrated even if the domain of m is restricted to eigenvectors of a large but still finite number of different sets $\{f_i\}$. Moreover these $\{f_i\}$ can be chosen to be eigenvector sets of operationally definable physical quantities. Hence, using KS instead of the weaker corollary (G)′ of Gleason's theorem, the above *reductio* can be repeated but

---

[3] I am here using 'eigenvalue' as synonomous with 'possible value'. On the more usual approach to QT the eigenvalue of Q is defined via the representative operator $\hat{Q}$ for Q, and it must then be advanced, as an independent assumption, that the spectrum of eigenvalues is the set of possible values.

without the need to assume (C). Instead we merely assume the existence of certain physical quantities the existence of which is guaranteed by QT in any case. This new *reductio* then enables us to derive $\sim$(U), instead of just $\sim$(U) *or* $\sim$(C), given (Det Q) and the laws of QT. (Note that the physical quantities required in this last proof, and which I claim QT assures us exist, include certain non-degenerate Q, Q' which are 'weakly incommensurable' in the sense that they do not share a complete set of eigenvectors.[4] This will be of importance later.)

So it seems that the realist interpreter of QT is forced to give up (U). I shall now consider the consequences of doing this. In giving up (U) we say that for any S, t (with implicit restriction to dimension $\mathscr{H}(S) \geq 3$, which I take for granted from now on) it is the case that

(i)   There are distinct non-degenerate physical quantities Q, Q' and a vector f in $\mathscr{H}(S)$ which is both the eigenvector of Q for some eigenvalue q_i and the eigenvector of Q' for some eigenvalue q'_j and $Q(S, t) = q_i$ but $Q'(S, t) \neq q'_j$.

Prima facie (i) may not seem problematic, but further consideration reveals otherwise. In particular we will now show that giving up (U) forces us to admit that the representation of physical reality offered by QT is 'incomplete' in a certain respect.

Assume that for any non-degenerate Q and any eigenvalue q_i of Q there is a q-quantity Q_i with possible values 1 and 0, for which:

(ii)   *Statistical coarse-graining condition.* If Prob (Q has value q_i in S at t) = 1 then Prob (Q_i has value 1 in S at t) = 1 and if Prob (Q has value q_j in S at t) = 1 for some $j \neq i$ then Prob (Q_i has value 0 in S at t) = 1,

and

(iii)   *Extensional coarse-graining.* If $Q(S, t) = q$ then $Q_i(S, t) = 1$ and if $Q(S, t) = q_j$ for some $j \neq i$ then $Q_i(S, t) = 0$.

Now from (i), (ii), and QT[5] we see that Q_i has as eigenvectors for eigenvalue 1 any eigenvectors of Q for eigenvalue q_i and these must

---

[4]   By contrast strong incommensurability requires *no* eigenvectors in common.

[5]   In particular from the law of QT which says that if Q is non-degenerate on $\mathscr{H}(S)$ then Q has value q_i in S at t with certainty iff $f(S, t) = f_i$, where f_i is the unique eigenvector of Q for eigenvalue q_i on $\mathscr{H}(S)$. Note that the place of this law in QT will be discussed in Chapter 9 where we will rework the above proof. For the moment I simply take it for granted.

include the vector f. And we similarly see from (ii) and QT that Q_i has as eigenvectors for eigenvalue 0 the set of vectors f' which are eigenvectors of Q for any eigenvalue q_j where $i \neq j$. Moreover f together with the latter set of vectors f' forms a complete set in $\mathcal{H}(S)$ (since the set of eigenvectors of Q in $\mathcal{H}(S)$ is complete). And we similarly derive that Q'_j has these same eigenvectors and eigenvalues in $\mathcal{H}(S)$. So we see that Q_i and Q'_j are identical from the point of view of their representation within $\mathcal{H}(S)$—they share a complete set of eigenvectors in $\mathcal{H}(S)$ and their corresponding eigenvalues are identical. (Indeed they are both represented by the projection operator $\hat{P}(f)$ onto f.) But from (iii) we see that $Q_i(S, t) = 1$ and $Q'_j(S, t) = 0$ since *ex hypothesi* $Q(S, t) = q_i$ and $Q'(S, t) \neq q'_j$. Hence Q_i and Q'_j cannot be the same physical quantity despite being represented identically within QT. In short they are distinct but not distinguishable within the formalism of QT.

This result faces us with one of two options. Either we accept the supposed existence of Q_i and Q'_j *qua bona fide* q-quantities, or we do not. If we follow the first option we must concede the 'incompleteness' of QT, i.e. we must concede (to paraphrase Einstein, Podolski, and Rosen) that there are distinct 'elements of physical reality', viz. Q_i and Q'_j, which are indistinguishable within the form of representation for those elements available within QT. (Note that here it is physical quantities, rather than physical quantities having particular values, which are the 'elements of reality', in contrast with EPR.) And in particular we must 'de-Ockhamize' the form of representation for physical quantities in QT, by allowing the one operator on $\mathcal{H}(S)$ to represent more than one physical quantity for S. What we have here called 'de-Ockhamization' is called 'ontological contextualization' by Heywood and Redhead (1983) p. 485; but I shall follow Shimony (1984) p. 29 in restricting the term 'contextualization' to those forms of representation for q-quantities where the *measured* values rather than the possessed values are allowed to depend on the context of measurement. (We shall discuss contextualization in the latter sense in section 3 of this chapter.)

The second option is to deny that the Q_i and Q'_j are q-quantities. And this either means (first sub-option) denying that they are proper physical quantities at all, or (second sub-option) admitting that QT is incomplete because the Q_i and Q'_j are proper physical quantities but not representable as q-quantities. Consider the first of these sub-options, in particular suppose there is no physical quantity satisfying

(ii) and (iii). But the existence of a Q_i satisfying (iii) is surely uncontroversial: if Q exists so does the extensionally coarse-grained Q_i satisfying (iii). (Although how, if at all, such a Q_i is to be represented within QT is another matter.) So what this second sub-option seems to reduce to is the claim that there is a Q_i satisfying (iii) which does not also satisfy (ii).

But the latter claim seems to be unsatisfactory for the following reason. Suppose there is a Q_i satisfying (iii). Then in any possible situation for which $Q(S, t) = q_j$ it will be the case that $Q_i(S, t) = 1$ or 0 depending on whether or not $i = j$, for any t. Now construct an ensemble of possible time-slices of the system S at times $t_1, t_2, \ldots$ all in the same state as S at t. For each $t = t_n$ (ii) holds; and hence the relative frequency with which $Q_i(S, t_n) = 1$ (or 0) over the ensemble (i.e. for varying n) is equal to the relative frequency with which $Q(S, t_n) = q_i$ over that ensemble. By making this ensemble large enough (and since the various members of the ensemble are statistically independent because each is in a pure state—as we shall discuss in the next chapter) we ensure that there is negligible probability that the relative frequency with which Q has a particular value over that ensemble differs from the probability with which Q has that value in S at t by any predetermined non-zero amount (by the law of large numbers). Thus although (ii) may fail, it is as unreasonable as one likes to take it as failing, i.e. the probability of it failing can be made as small as we like. As such the sub-option of denying that there is a proper physical quantity satisfying both (ii) and (iii) turns out to be unacceptable, unacceptable because the extensional coarse-graining condition for Q, viz. (iii), is uncontroversial and (as we have just shown) by accepting (iii) we are also it seems committed to the statistical coarse-graining condition for Q, viz. (ii).

In short KS shows that if we adopt the realist programme of interpreting QT within a metaphysical framework which includes (Det Q) then our only acceptable options are either to adopt the de-Ockhamizing strategy, and *a fortiori* admit the incompleteness of QT, or admit the incompleteness of QT by admitting that there are physical quantities, viz. Q_i, which are not representable as q-quantities. This conclusion agrees with EPR's conclusion that QT is incomplete. But the argument just presented for this conclusion, based on KS, avoids various of the controversial aspects of EPR's argument; in particular it does not rely on EPR's vexed criterion for being an 'element of physical reality'.

This result has implications not just for realist interpretations of QT which endorse (Det Q), but also for the interpretation which I have been putting forward here. According to the principle (Bohr)' of Chapter 4 if $\hat{W}(S, t)$ is the identity operator then all q-quantities for S at t have a determinate value. So, in this special case at least, the interpretation I have been putting forward agrees with (Det Q), and so, given the results of the preceding argument, must admit the incompleteness of QT. (Although of course this incompleteness will only be manifested in very special cases, viz. those for which $\hat{W}(S,t)$ is the identity operator.)

2. DERIVING INCOMPLETENESS (B)

There is an alternative approach to demonstrating that the incompleteness of QT is implicit in a commitment to (Det Q) within the context of QT. To explain this alternative, and more usual, approach we must introduce an important idea. Let F be any function from numbers onto numbers. We then define '$F(Q)$', where Q is a physical quantity (*not* a number), q.v.:

> $F(Q)$ is the physical quantity for which for any possible S,t the value of $F(Q)$ in S at t is just $F(Q(S, t))$.

In this way the notion of a function is extended from a function of the numerical values of physical quantities to the physical quantities themselves. Note however that $\hat{Q}' = F(\hat{Q})$[6]—a functional relation between the representative operators $\hat{Q}'$ and $\hat{Q}$ for the physical quantities Q and Q' respectively—does not guarantee that $Q' = F(Q)$ in the sense that $Q'(S,t) = F(Q(S,t))$ for all actual S, t. Indeed:

(Func) $Q' = F(Q)$ if $\hat{Q}' = F(\hat{Q})$

will not be one of the assumptions which we introduce on this alternative approach. We do however, introduce the assumptions:

(Func)$_1$ If Q is a q-quantity then so is $F(Q)$ *and* $\hat{F}(Q)$, the operator representing $F(Q)$, $= F(\hat{Q})$.

(U)$_1$ If $\hat{Q} = \hat{Q}'$ then $Q = Q'$; i.e. $Q(S, t) = Q'(S, t)$ for all actual S, t.

---

[6] The notion of a function of operators was defined above: $F(\hat{Q})$ is an operator with eigenvector f for eigenvalue $F(q)$ if $\hat{Q}$ has eigenvector f for eigenvalue q, for all such f. Note however that this only completely specifies $F(\hat{Q})$ if $\hat{Q}$ has a totally discrete spectral resolution. I restrict myself to considering just such operators from now on.

It is then easy to show that $(\text{Func})_1$ entails the existence of the coarse-grained physical quantities Q_i discussed in section 1, as follows.

Let F be the function D_i defined by:

$$D_i(q) = 1 \text{ if } q = q_i$$
$$\qquad = 0 \text{ if } q = q_j \text{ for any } j \neq i.$$

Then clearly $D_i(Q)$ is a q-quantity (by $(\text{Func})_1$) and by definition of D_i satisfies:

$$D_i(Q)(S,t) = 1 \text{ if } Q(S,t) = q_i$$
$$D_i(Q)(S,t) = 0 \text{ if } Q(S,t) = q_j \text{ for } j \neq i.$$

Also, since $\hat{D}_i(Q) = D_i(\hat{Q})$ (by Func_1), it follows that not only does $D_i(Q)$ have the eigenvector f for eigenvalue 1, but it also has any f', orthogonal to f, as eigenvector for eigenvalue 0. Thus $D_i(Q)$ satisfies:

Prob $(D_i(Q)$ has value 1 or 0 respectively in S at $t) = 1$ iff Prob $(Q$ has value q_i or q_j respectively in S at $t) = 1$ for any $j \neq i$.

In short $D_i(Q)$ satisfies exactly the conditions on Q_i given in (ii) and (iii). Thus (Func) entails the existence of Q_i. QED.

Moreover it is trivial to derive (U) from $(\text{Func})_1$ and $(\text{U})_1$. Let the eigenvector of Q for eigenvalue q_i also be the eigenvector of Q' for eigenvalue q'_j. Then Q_i and Q'_j both exist as q-quantities (from $(\text{Func})_1$); and clearly $\hat{Q}_i = \hat{Q}'_j$. Hence, by $(\text{U})_1$, $Q_i = Q'_j$ which is just what (U) says. QED. Thus we see that instead of the argument of section 1 starting from (U) and the existence of the coarse-grained physical quantities we could have started from the stronger $(\text{Func})_1$ and $(\text{U})_1$; or indeed from an even stronger combination consisting of $(\text{Func})_1$ and (Func). ((Func) entails $(\text{U})_1$ trivially, by letting $\hat{F}$ be the identity operator.) This approach is the one followed by Healey in (1979) for example.

I shall now draw out some further consequences of the above result. Assume that the Q_i described above happens to be measured for S at t. Standard QMT (see principles $(\text{U})_0$ and $(\text{U})'_0$ of Chapter 5) assures us that, since Q_i and Q'_j have the same representative operators, then a measurement of Q_i is *a fortiori* a measurement of Q'_j and vice versa, so that the assumption here means that not only Q_i but equally Q'_j are measured. Moreover QMT assures us that the measured values of Q_i and Q'_j for this joint measurement of both Q_i and Q'_j are the same (see

(Cal)' in Chapter 5):

$$\text{mv}(Q_i, S, t) = \text{mv}(Q'_j, S, t).$$

(These mvs are *not* counterfactual, since Q_i and Q'_j are taken to be actually measured.) But because

$$Q_i(S, t) \neq Q'_j(S, t)$$

it follows that either $\text{mv}(Q_i, S, t) \neq Q_i(S, t)$ or $\text{mv}(Q'_j, S, t) \neq Q'_j(S, t)$; and hence another of the realist's principles, viz. (Pass Q), fails.

The realist has several alternatives to accepting this last conclusion. Firstly he could deny that the Q_i and Q'_j are ever measured. But this would seem to require a far-reaching conspiracy on the part of nature in order to ensure that whatever S, t are, we never get to measure that particular Q_i for which there is a Q'_j (with the same representative operator) such that $Q_i(S,t) \neq Q'_j(S,t)$. A more plausible alternative is to revise standard QMT by restricting the proper measurement of a *q*-quantity to the measurement of some non-degenerate physical quantity of which the quantity to be measured is a function, and then appeal to:

(Func)$_0$ $\text{mv}(F(Q), S, t) = F(\text{mv}(Q, S, t))$ for non-degenerate Q.

We will denote the $\text{mv}(F(Q), S, t)$ defined in this way by '$\text{mv}_Q(F(Q), S, t)$'. (This alternative is the one proposed by van Frassen (1973) p. 108 as the 'anti-Copenhagen' variation.)

The restriction of the class of ideal measurements achieved by this move is severe. For example the following usual method of measuring a physical quantity $\bar{Q}$ represented by a projection operator $\hat{P}(f)$ is ruled out. The usual method is that of calibrating a measuring apparatus so as to register the value 1 whenever S at t is in the state f and to register the value 0 when S at t is in any state orthogonal to f. But this method of measurement is no longer allowed because it does not measure a non-degenerate Q. Instead we must measure Q via measuring some non-degenerate Q for which different apparatus readings have been correlated with each of the eigenvectors of Q, one of those eigenvectors, say f_i, being f. And even then we are not entitled to conclude from '$\text{mv}(Q, S, t) = q_i$' to '$\text{mv}(\bar{Q}, S, t) = 1$' unless $\bar{Q}$ happens to be a function of Q. For the realist however there is a high return for making this severe restriction on the class of measurements, viz. the preservation of (Pass Q).

3. MEASURED VALUES

KS can however be turned to more account than just showing that from (Det Q) as a premiss we can derive the incompleteness of QT. In this section I will demonstrate some of these extra resources which KS has, resources which do not depend on taking (Det Q) as a premiss. In particular I shall show that if we assume, instead of (Det Q), that

(MV)   For every q-quantity Q (de-Ockhamized or not) and for any q-system S at any time t (during the lifetime of S) there is a unique measured value, counterfactually construed, which is the value which Q would be measured to have were it measured, and which we denote by '$mv(Q, S, t)$'.

then we are forced to 'contextualize' mvs. (Note that (MV) pre-supposes that for every Q,S,t, Q is measurable in S at t, since otherwise the conditional 'Were Q measured in S at t then it would be measured to have value q' would be true for any q, and hence there would be several such counterfactual measured values for Q in S at t.)

Let f_i be the eigenvector of the non-degenerate q-quantity Q for S, for eigenvalue q_i. We construct a map m' from the set of such f_i in $\mathscr{H}(S)$ onto 1 or 0 q.v. We associate 1 with f_i if $mv(Q, S, t) = q_i$ and associate 0 with f_i if $mv(Q, S, t) = q_j$ for $q_j \neq q_i$. And we can extend this to a single-valued map m' over a number of such sets of eigenvectors (for different Q) if we assume the uniqueness condition:

(U)'   If f is the unique eigenvector of Q for eigenvalue q and of Q' for eigenvalue q' then $mv(Q, S, t) = q$ iff $mv(Q', S, t) = q'$.

But the existence of such an m' is not possible according to KS (see section 1), or at least this is so when (as will be the case for some S) $\mathscr{H}(S)$ is more than three dimensional and the domain of m' is extended over a sufficiently large number of eigenvector sets (sufficiently large to include totally disjoint complete sets of eigenvectors). So KS forces us to give up (U)' if we are to retain (MV). And this in turn means accepting:

(iv)   For some S, t there exist Q,Q' and an f for which f is the eigenvector of Q for eigenvalue q and also the eigenvector of Q' for eigenvalue q' and $mv(Q, S, t) = q$ but $mv(Q', S, t) \neq q'$.

(The necessity of giving up (U)' if we insist on (MV) is just like the necessity of giving up (U) if we preserve (Det Q), as discussed in section 1. In both cases it is KS which underwrites the necessity, but in the former case it is possessed values which we consider whereas in the present case it is measured values.)

We can now distinguish two sub-options depending on whether or not we retain (Det Q). First suppose we reject (Det Q) thus obviating the need to de-Ockhamize. In such a case we surely ought not to de-Ockhamize (since the onus of justification lies with the de-Ockhamizer). And if we follow this prescription then we can define a unique physical quantity $\bar{Q}$ which is the physical quantity represented by $\hat{P}(f)$. But then from (iv) and (Func)$_0$ we see that $mv_Q(\bar{Q}, S, t) \neq mv_{Q'}(\bar{Q}, S, t)$; and hence we must also reject the idea that there is always a unique measured value for $\bar{Q}$ which is independent of how $\bar{Q}$ is measured. In other words we see that the measured values must be 'contextualized' for the following reason: (MV) and (Func)$_0$ imply that there is a unique mv for $\bar{Q}$ in S at t when $\bar{Q}$ is measured by measuring Q (or at least by measuring a non-degenerate q-quantity of which Q is a function), and which we denote by '$mv_Q(\bar{Q}, S, t)$'. And similarly there is a unique $mv_{Q'}(\bar{Q}, S, t)$. But we have $mv_{Q'}(\bar{Q}, S, t) \neq mv_Q(\bar{Q}, S, t)$ for some $\bar{Q}, Q, Q'$; and hence there is no value for a particular method of measuring $\bar{Q}$ in S at t which is independent of how $\bar{Q}$ is measured. It is this latter feature, viz. that $mv\bar{Q}$ depends on the context of measurement (i.e. on how $\bar{Q}$ is measured), which we refer to as the 'contextualization of mvs'. Again we have a revision of standard QMT (as we did at the end of section 2).

It is easy to see that the contextualization referred to here is precisely what Shimony calls 'algebraic contextualization', or (adapting Heywood and Redhead's term from another context)[7] 'ontological contextualization'. That is 'contextualization', as I use that term here, refers to a dependence of the measured value for $\bar{Q}$ on the method of measuring $\bar{Q}$; in particular it refers to the fact that $mv_Q(\bar{Q}) \neq mv_{Q'}(\bar{Q})$ for Q, Q' with different sets of eigenvectors. Hence there is a dependence of $mv_Q\bar{Q}$, via its dependence on Q, on the complete orthonormal set of eigenvectors of Q, and it is just this latter dependence which Shimony, following Gudder, calls 'algebraic contextualization'.

Note however that notwithstanding $mv_{Q'}(\bar{Q}, S, t) \neq mv_Q(\bar{Q}, S, t)$ there may still be a counterfactual $mv(\bar{Q}, S, t)$. This will trivially be so

---

[7] As discussed in sect. 1 of this chapter, Heywood and Redhead use the term 'ontological contextualization' for what we call 'de-Ockhamization'.

in the special case that $\bar{Q}$ is in fact measured in S at t (by whatever method), because mv($\bar{Q}$, S, t) will then simply be the value reported by the measurement of $\bar{Q}$ which in fact takes place in S at t.[8] Moreover, even if no measurement of $\bar{Q}$ actually takes place in S at t, mv($\bar{Q}$,S,t) may exist, despite mv$_{Q'}$($\bar{Q}$, S, t) $\neq$ mv$_Q$(Q, S, t). For example this will be so if there is some one method of measuring $\bar{Q}$, by measuring Q', say, which is 'closer to realization' than any other.[9] More particularly we can assume—and from now on this is what I shall call 'the contextualization strategy'—that there is a unique mv(Q, S, t) for any Q (degenerate or not), but that a degenerate $\bar{Q}$ may nevertheless have different mvs if its measurements are restricted ('contextualized') to measurements of different non-degenerate Q of which $\bar{Q}$ is a function.

The other sub-option is to keep (Det Q) and hence to de-Ockhamize. If we do this then we do not need to contextualize, but we are back with the unattractive options of section 1, and in particular the restriction on measurements implicit in de-Ockhamization.

4. BOHR'S REJECTION OF MEASURED VALUES

The argument of the previous section required the coexistence of mv(Q, S, t) and mv(Q', S, t) for Q, Q' strongly incommensurable. This is because, in order for KS to rule out the existence of m', m' must include in its domain at least two totally disjoint complete sets of eigenvectors. Hence the conclusion of the previous section, viz. that we need either to contextualize or return to the de-Ockhamization of section 1, can be avoided if we agree with Bohr's view that counterfactual measured values for both Q and Q' in S at t cannot exist if Q and Q' are strongly incommensurable. With one minor qualification I shall argue in favour of Bohr's view here, but let me firstly criticize those Copenhagan interpreters (and perhaps Bohr himself) who would defend what I have here called 'Bohr's view' along the following verificationist lines. First they would claim that if Q and Q' are 'incommensurable' in the sense that they share no eigenvectors then they cannot be simultaneously measured. For

[8] This is on a Lewis view of counterfactuals for which '$p \rightarrow q$' is true if '$p.q$' is true—see Lewis (1973), and later discussions here.
[9] This is on a Lewis view of counterfactuals for which '$p \rightarrow q$' is true if all nearest p-worlds to the actual are q-worlds.

argument's sake let us accept this first claim.[10] On the basis that Q and Q' cannot be measured simultaneously they can go on to claim that it is impossible to determine both mv(Q, S, t) and mv(Q', S, t)— presumably because it is impossible to realize the antecedents of both counterfactuals of the form 'Were Q measured in S at t the mv would be. . .' and 'Were Q' measured in S at t the mv would be. . .' if Q and Q' cannot be measured simultaneously. And from this they then conclude—here the verificationism enters—that it is meaningless to assign values to both mv(Q, S, t) and mv(Q', S, t). That is, what cannot be determined cannot be meaningfully asserted. In short both these mvs cannot exist.

It is easy to criticize this defence of Bohr's views. Not only does it share the bankruptcy of verificationism, but it also ignores the fact that we do have a means of access to counterfactuals other than by realizing their antecedents. Consider a piece of glass. We may well know by direct testing that it would break under a 100-lb. weight (viz. by seeing that it does in fact break when a 100-lb. weight is placed on it); but we will then also know indirectly that it would have (counterfactually) broken under a 200-lb. weight. This additional knowledge is derived from the fact that the glass did break under the 100-lb. weight together with knowledge of the causal regularity that any piece of glass has a certain breaking stress which whenever it is exceeded causes it to break. Thus, *contra* the verificationist, I can have good grounds for asserting both 'Were I to place a 100-lb. weight only on this piece of glass it would break' as well as 'Were I to place a 200-lb. weight only on this piece of glass it would break' even though the antecedents of these two counterfactuals are not compossible.

So what is there to be said in favour of Bohr's view here (his view that the strongly incommensurable Q and Q' cannot both be assigned mvs in S at t) if the verificationist defence of that view is invalid? The answer lies in another metaphysical assumption made by Bohr, viz. the assumption that the outcomes of measurement in QT are *indeterministic* when the measured system is in a pure state which is not an eigenstate of the measured physical quantity. (Cf. Bohr's quite different claim, discussed in Chapter 1, that a physical quantity is *indeterminate* in value when the system under consideration is in a pure state which is not an eigenstate of the physical quantity.

[10] Although the usual justification for this claim depends on a somewhat dubious account of measurement. See Jammer (1974), p. 83 for a discussion of simultaneous measurements.

Indeterminacy *qua* fuzziness clearly needs to be distinguished from indeterminism.) I shall now propose a rough analysis of what is meant by 'indeterministic' here, and then explicate it in terms of the possible world framework which Lewis uses to explicate counterfactuals. This explication will enable me to defend Bohr's view given at the beginning of this paragraph but without appeal to verificationism.

The Lewis analysis of counterfactuals (as in Lewis (1973) p. 17) presupposes the existence of various possible worlds as well as an ordering relation between possible worlds which locates some of the possible worlds as nearer to the actual world than others. In particular Lewis supposes the actual world to be surrounded by nested spheres of possible worlds such that the world w, which belongs to a sphere S which is inside the sphere S', is nearer to the actual world than any of those worlds belonging to the sphere S' which do not also belong to the sphere S. Each possible world w—or at least each world w accessible from the actual world—is taken to belong to some such sphere S (and of course belongs to all spheres outside of S as well). I shall use the locution 'the P-worlds within S' to abbreviate 'those possible worlds at which P is true and which belong to the sphere of possible worlds S'. Note that 'within' is here only meant topologically: there is no assumption of a metric over the space of possible worlds. Nor is there need to specify a comparative nearness relation on all triples of worlds: to specify whether or not w is nearer to w' than w'', for all worlds w, w', w''. Note also that for convenience I shall take the term 'possible world' in what follows to refer only to a possible world which is accessible from the actual, i.e. which falls within the set of spheres centred on the actual world. And I shall use the sign $\leq$ to stand both for the relation of being nearer to the actual (as among possible worlds) as well as for the relation of being a proper subset of (as applied to the nested spheres of possible worlds about the actual).

Lewis distinguishes two cases in which the counterfactual (or more broadly subjunctive) conditional 'Were P then Q'—denoted '$P \rightarrow Q$'—is warrantedly assertible ('true'). First of all, in the case that there are some P-worlds, '$P \rightarrow Q$' is true iff there is a sphere S of possible worlds for which there are P-worlds within S, and any such P-worlds are also Q-worlds. The second case is the vacuous case where there are no P-worlds, and in that case '$P \rightarrow Q$' is simply legislated to be true.

Now to say that the outcome R of some particular (but perhaps merely hypothetical) set-up E is indeterministic means (roughly) that the outcome is not fixed by the set-up E, i.e. that over a series of re-runs of E the outcome would vary—sometimes R, sometimes not—and, more precisely, that E does not lead to the same result in all possible situations near enough to the actual at which E is repeated. Explicating this condition in Lewis's terminology suggests that 'R is an indeterministic outcome of E' means that, for any set of near-enough possible E-worlds, R occurs in some but not all of them. Note however that on a Lewis account of counterfactuals the latter explication is only useful if E does not take place in the actual world, since if it does take place there then the actual world is the unique nearest possible E-world, and so either all nearest E-worlds are R-worlds or none of them is.

If we accept this explication then we see that 'R is an indeterministic outcome of E' implies the falsity of both the counterfactuals 'Were E to take place then R would occur' and 'Were E to take place then R would not occur', at least it does so for E non-actual. So now consider the statement 'Were Q measured in S at t the measured result would be q'. Clearly from what we have just said this too is false, and is false for all q, if the various possible outcomes (i.e. whether or not the measured value is q_i) of the Q-measurement are indeterministic (for all i), *and* if no measurement in fact takes place, i.e. if the measurement is 'counterfactual' in the strict sense. In short, there is no counter-factually construed (in the strict sense) measured value for Q in S at t if the various outcomes of the measurement process are indeterministic.

Note that the sense of 'indeterminism' introduced here is somewhat non-standard in that it does not construe indeterminism and determinism (in one common sense of the latter term) as exclusive. This is because the results of certain hypothetical repeat measure-ments may be fixed (determined) by the various factors which characterize the various repeat measurement situations, and yet there may be different measured results (because of slightly different causal antecedents) for two equally nearest possible repeats. In that case we have both 'determinism' (in the sense that the results are fixed by their causal antecedents) and indeterminism (in the sense that there is no counterfactually construed mv). This point will be of importance shortly. It indicates that even for 'deterministic' measurements (MV) may fail, i.e. we may have indeterminism.

We are now in a position to draw some consequences from Bohr's (and the Copenhagen interpretation's) assumption that the outcomes of a measurement process are indeterministic when the measured system is in a pure state which is not an eigenstate of the measured physical quantity. We see that it means that mv(Q, S, t) does not exist if $f(S, t)$, the state-vector for S at t, is not an eigenstate of S at t, and if the measurement of Q in S at t is strictly counterfactual. And it is further easy to prove that this result justifies in qualified form what we initially postulated, as part of Bohr's view, concerning the non coexistence of mvs for strongly incommensurable physical quantities, and hence the failure of all mvs to coexist. In particular we can prove that, at least for pure states, and if the measurements of Q in S at t are all strictly counterfactual (i.e. no measurements actually take place in S at t) then the existence of mv(Q, S, t) precludes the existence of mv(Q', S, t) if Q and Q' are incommensurable in the strong sense that they share no eigenvectors at all:

Proof Either $f(S, t)$ is an eigenvector of Q or of Q', but not of both (since Q and Q' are strongly incommensurable). Hence either the outcome of the measurement of Q is indeterministic, or the outcome of the measurement of Q' is indeterministic. Hence the strictly counter-factual mv(Q, S, t) does not exist, or the strictly counterfactual mv(Q', S, t) does not exist. QED.

Note however that if one of the strongly incommensurable Q' and Q is in fact measured then mv(Q, S, t) trivially exists—let it be Q—but the counterfactually construed mv(Q', S, t) may then also exist if $f(S, t)$ happens to be an eigenstate of Q'. Thus mv(Q, S, t) and mv(Q', S, t) may coexist despite Q and Q' being strongly incommensurable but only if one of them is not *strictly* counterfactual. Note also that the previous proof does not work if Q and Q' are merely *weakly* incommensurable since they may then share some eigenvectors (albeit not a complete set). Thus weakly incommensurable physical quantities may have simultaneous mvs on the Bohr view, as I have construed it here.

We have proven that (MV) fails, and more particularly that (the strictly counterfactual) mv(Q, S, t) and mv(Q', S, t) cannot both exist if Q, Q' are strongly incommensurable. This in turn relieves us of the need to contextualize measured values notwithstanding the argument of the previous section, since that argument was premised on an assumption of (MV). These conclusions do however presuppose taking seriously Bohr's views of the outcomes of a Q-measurement on

S at t as indeterministic if S at t is in a pure state which is not an eigenvector of Q.

It is important to note that indeterminism, as I have construed it here, namely as what Stapp would call 'failure of counterfactual definiteness',[11] does not imply randomness or the existence of intrinsic probabilities. A more complete account of the relation of indeterminism in this latter sense to indeterminism in the more usual sense of that term, for which indeterminism means the existence of intrinsic probabilities, is the subject of the next section. There I shall show that indeterminism *qua* the failure of counterfactual definiteness is implied by indeterminism *qua* the existence of intrinsic probabilities, *if* we first make an obvious extension of Lewis's truth-conditions for counterfactuals. (This same extension will turn out to enable an extension of the above definition of indeterminism to cover non-actual processes.)

This result which I shall derive in the next section may appear to reverse the endorsement which we gave earlier (in Chapter 3) to Mellor's claim that the existence of intrinsic probabilities is consistent with determinism. But it is only if we see determinism and in-determinism as mutually exclusive that we are committed to such a reversal. And it is precisely that which we deny here. To put matters in terms of the coin-tossing example discussed in Chapter 3—the hypothetical tossing of a coin may be deterministic in that its outcome is causally connected with certain initial conditions; and yet the outcome may be indeterministic in the sense just given here because minute possible variations in the initial conditions (variations which do not remove the toss any further from the actual world than it is) cause a change in the outcome of the toss.

5. INTRINSIC PROBABILITIES AND MEASURED VALUES

Bohr's assumption of indeterminism does allow us to justify the failure of (MV), but does not give us much insight into why it occurs. Indeed the assumption of indeterminism amounts to little more than a reassertion of the failure of (MV). I shall rectify this lack of insight in this section by showing how indeterminism, and more particularly the failure of (MV), emerges from the existence of an intrinsic

[11] See Stapp (1977).

probability distribution over possible outcomes for measurement. (Note that implicit in this last claim is the view that the existence of intrinsic probabilities is consistent with determinism, *contra* Mellor's claims in (1971), but in line with his later (1982).)

In putting forward these views I am extending Redhead's (1983) and Hellman's (1982b) views. Both of them agree that it is the 'ultimately random' (Hellman (1982b) p. 491) or 'essentially probabilistic' (Redhead (1983) pp. 161 and 162) nature of measurement which leads to the failure of (MV). But neither shows how this connection can be made.[12] Moreover neither clearly distinguishes indeterminism *qua* the failure of counterfactual definiteness from the existence of intrinsic probabilities. By contrast I shall argue that there is an important distinction to be made between the existence of intrinsic probabilities and indeterminism, and give a sketch of how the former can be seen to entail the latter and more particularly the failure of (MV). More generally I shall argue here that the failure of (MV) arises not from a failure of determinism, in particular not from an acausal 'collapse of the wave-packet', but rather from the fact that if $f(S, t)$ is not an eigenvector of Q then there is a non-trivial intrinsic probability distribution over the various possible outcomes of measurement of Q on S at t ('non-trivial' means that none of the probabilities has value 1).

I shall also point out that the existence of intrinsic probabilities is quite consistent with determinism (which, as we indicated at the end of the section 4, is not excluded by indeterminism as we are using those terms). This is an important point to make because if on the contrary one were to tie the failure of (MV) to the failure of determinism (rather than the existence of intrinsic probabilities) then deterministic hidden variables (hv) theories would not be able to admit the failure of (MV), and hence, in the light of the argument of section 3, would be forced to contextualize (algebraic contextualization) their mvs. It will be a result of my arguments here that, on the contrary, (MV) may fail for deterministic hv theories, and hence we see that the argument of section 3 no longer imposes the need to contextualize mvs. On the other hand the argument of section 1 will still show that such deterministic hv theories must, in virtue of

---

[12] Redhead merely gives the explanation I advanced above for the failure of (MV) in terms of indeterminism, and Hellman only makes the tentative claim that if we remove determinism then (MV) becomes 'highly suspect'.

endorsing (Det Q)—as do all hv theories—provide a de-Ockhamized representation for q-quantities, even though their mvs do not need to be algebraically contextualized (because they allow the failure of (MV)).

Consider a Q-measurement process in which the combined state of the final measuring apparatus (S_2) and measured system (S_1) takes the familiar correlated form

$$\Sigma c_i f_i^1 \times g_i^2$$

and the initial state of S_1 is

$$\Sigma c_i f_i^1.$$

Then $\hat{W}_2$, the final measuring apparatus density operator, is given by:

$$\Sigma |c_i|^2 \hat{P}(g_i^2),$$

where g_i^2 corresponds to a state in which the apparatus registers q_i so that there is probability $|c_i|^2$ of the measured value being q_i. And the probability here is objective, single-case, or more briefly, 'intrinsic'. (The alternatives of taking the probability as subjective, or a frequency, lead to difficulties which we have already discussed in Chapter 3, p. 53 ff and Chapter 4, p. 95 ff. Eg the frequency view, manifested in the ignorance interpretation of mixtures, cannot explain how the behaviour of $S_1 + S_2$ which is in the pure state $\Sigma c_i f_i^1 \times g_i^2$ differs objectively and detectably from how it would behave were it in the mixed state $\Sigma |c_i|^2 \hat{P}(f_i^1) \times \hat{P}(g_i^2)$.)

Now suppose that the initial pure state of the measured system is not an eigenvector of Q, so that $|c_i|^2 < 1$ for all i, i.e. there is a non-trivial intrinsic probability distribution over the various possible measured results. The question then is why should this non-trivial intrinsic probability distribution over various possible measurement results mean that the outcome of the measurement process is indeterministic in the modal sense defined above? I can only sketch a reply to this question since it depends on the complex issue of how objective single-case probabilities relate to singular modal relations. My suggestion is to define 'there is an objective, single-case probability p of a particular experiment E having the result R' to mean something like the following. Suppose that the topology of the space of possible worlds orders the spheres of possible worlds (about the actual) into various 'monotone', i.e. nested, sequences of spheres $\{S_n\}$ for which $S_n < S_m$ for any $n < m$, and which are 'complete' in the sense that for any S there is an S_n for which $S_n < S$. We then distinguish three cases.

In the first case there is some complete set $\{S_n\}$, and some continuous measure m on the sets of possible worlds, for which $m(E.S_n) \neq 0$ for *all* n. (Here '$E.S_n$' is the set of possible worlds within S_n at which a counterpart of the particular experiment E occurs.) In that case we assume

(a) $P(E$ has result $R)$ if it exists $= \text{limit}_{n \to \infty} m(E.R.S_n)/m(E.S_n)$, for some such $\{S_n\}$ and some continuous m.

The second case is one for which there are no $\{S_n\}$ as described in the first case and moreover, for some continuous m, and for some complete $\{S_n\}$, there is an $\bar{S}$ which is a least upper bound on those spheres S_n for which $m(E.S_n) = 0$, i.e. for any S_n, where $S_n > \bar{S}$, $m(E.S_n) > 0$. In that case we assume

(b) $P(E$ has result $R)$ if it exists $= \text{limit}_{S_n \to \bar{S}} m(E.R.S_n)/m(E.S_n)$ for some such $\{S_n\}$, and some continuous m.

The third case is one for which no $\{S_n\}$ as described in the first case exist, but also there is no least upper bound $\bar{S}$ as described in the second case for any continuous measure m or any $\{S_n\}$, so that for any continuous m, $m(E.S_n) = 0$ for any S_n; and in particular (letting $S_n =$ the whole space) $m(E) = 0$. In this case we take it that $P(E$ has result $R)$ does not exist. These three cases are clearly exhaustive and exclusive, and in that sense provide a complete definition of $P(E$ has result $R)$. Note that the singleton set consisting of just the actual world must not count as a measurable set here otherwise the above definition implies that, for E actual, $P(E$ has $R)$ is trivially 1 or 0 (depending on whether E actually has R or not). Note also that the set of possible worlds here must have infinite cardinality if the probabilities are to be not just rational numbers.

Now in Appendix 10 I show that, on this account of single-case objective probability, the following theorem holds:

Theorem   If $P(E$ has $R) < 1$ then, for any S which contains a non-zero measure of E worlds (the measure as defined in (a) or (b)), some non-zero measure of the E-worlds in S are R-worlds.

From this it follows trivially that if $P(E$ has $R) < 1$ then '$E \to R$' is false, provided we make an obvious extension of the Lewis truth-conditions for counterfactuals:

> '$p \rightarrow q$' is true iff *either* the set of p-worlds has measure zero (trivial case) *or* for some sphere of worlds S there are spheres of worlds S' within S where each such S' contains a non-zero measure of p-worlds *and* all but a non-zero measure of those p-worlds in S' are also q-worlds.

Note that the suggestions here also complete a programme started in Chapter 2 for interpreting objective single-case probabilities. This interpretation captures the contingent nature of these probabilities, i.e. that they are not a priori probabilities, since the values of the probabilities are made to depend on the (contingent) location of the actual world in the space of possible worlds. On the other hand the probability values are not just given by limiting relative frequencies in the actual world, since they also depend on the structure of the space of possible worlds in the vicinity of the actual world. The plausibility of this interpretation for objective single-case probabilities does nevertheless derive from a generalized version of the frequency interpretation for which the probability of a particular trial E having a result of the type R is analytically connected with the relative frequency with which R occurs over a *hypothetical* ensemble of repeats of E.[13] What the suggestion here does is simply replace the hypothetical ensemble of 'repeats' of trial E with an ensemble of different possible worlds in each of which there is a counterpart of E (in Lewis's sense of counterpart), and also replaces the relative frequency with a relative measure. Note that it is easy to prove (see Proof 2 in the Appendix 10) that Lewis's 'Principal Principle' emerges as a consequence of this interpretation. It is this sort of consequence which indicates that the interpretation of objective single-case probabilities suggested here is not just *ad hoc*, i.e. it is not just suggested in order to sustain the claimed connection between the existence of intrinsic probabilities and indeterminism.

Now from the above result if $P(E$ has $R) < 1$ then '$E \rightarrow R$' is false, it follows immediately that if there is a non-trivial probability distribution over the possible results of measurement of Q on S at t, as there is when $f(S, t)$ is not an eigenvector of Q, then there is no one value of

---

[13] Popper can be seen as endorsing a generalized frequency interpretation of this form (1983b), pt. 2. Also see van Frassen's suggestive explication of this interpretation in his (1980) p. 190.

which it may be said that the measurement of Q in S at t would register that value. Thus we have indeed shown that just from the intrinsically probabilistic nature of QT it follows that (MV) fails. Moreover we see that even for a deterministic hv version of QT, (MV) fails, as long as we retain the intrinsic probabilities. (That we can do just that, i.e. combine determinism with objective chance, is one of the claims made above, following Mellor (1982). Note that, as indicated above, it is implicit here that we do not simply take determinism and indeterminism *qua* failure of counterfactual definiteness, as exclusive.)

6. MORE PASSIVITY PRINCIPLES

The above considerations based on KS have pointed to ambiguities in the principle (Pass Q) which I shall now discuss. In saying, with (Pass Q), that 'measured values = possessed values' there are two ambiguities to be resolved. There is the question of whether this identity is merely conditional on both measured and possessed values existing, or whether the identity is to be taken as additionally implying that if one of these values exists then so does the other. (Note that in what follows I implicitly take 'measured value' as counterfactual, or at least subjunctive.) And there is the further question of how to accommodate the fact that measurements of the one physical quantity may be carried out by any one of several different methods. For example is the above identity to be taken as asserting merely that there is some particular method of measurement which is faithful, or is it to be generalized over all methods of measurement, or then again is it to be taken as asserting merely that the measurement is faithful for the method of measurement which would be used were a measurement to be performed?

These considerations lead one to distinguish three groupings of passivity principles: (Pass Q)$_1$, (Pass Q)$_2$, and (Pass Q)$'_2$. (Pass Q)$_1$ principles assert the equality of measured values and possessed values conditional on the relevant possessed and measured values existing. The corresponding (Pass Q)$_2$ principles assert more strongly that when there is a possessed value then there is an equal measured value; and (Pass Q)$'_2$ principles (also stronger than the corresponding (Pass Q)$_1$ principles) assert the converse, that when there is a measured value there is an equal possessed value. (It is implicit here that the measurements are ideal.)

Within each of the these three groups of principles we can then distinguish three separate types of principles: a principle for which 'the measured value' means 'the value which would be reported in common by all methods of measurement', an 'unqualified' principle in which 'the measured value' means simply 'the value which would be reported were a measurement performed', and the most restricted principle in which 'the measured value' is understood as 'some one of the values which would be reported by some method of measurement'. We denote these three types of principles respectively by '(Pass Q)', '(U-Pass Q)', and '(R-Pass Q)'. For example '(R-Pass Q)$_2$' denotes the (R-Pass Q) type of principle in the second group of principles, and says that if $Q(S, t)$ exists then, for some method M of measuring Q in S at t, $mv_M(Q, S, t) = Q(S, t)$, etc. (A complete list of these principles appears in the Index of Principles.)

The point of making these distinctions will become apparent in the concluding section of this chapter where we discuss how these various passivity principles can be combined with (Det Q) in the light of the arguments of the preceding sections. For the moment, however, I point out some of the logical relations between these passivity principles. In particular I shall prove that although in general (Pass Q)$_2$ implies (U-Pass Q)$_1$ (which in turn is weaker than (U-Pass Q)$_2$), (Pass Q)$_2$ only implies (U-Pass Q)$_2$ in a special case; in particular we have:

Theorem 1   (Pass Q)$_2$ implies (U-Pass Q)$_1$

and

Theorem 2   (Pass Q)$_2$ implies (U-Pass Q)$_2$ if there is only a finite number of methods of measurement.

Moreover I shall prove that in general

Theorem 3   (U-Pass Q)$_2$ implies (R-Pass Q)$_2$.

The non-technical reader may skip these proofs, and now proceed to the conclusion of this chapter. The proof of these theorems requires the proof of certain lemmata:

Lemma 1   If $mv(Q, S, t)$ exists $= q$ then for some method of measurement M, $mv_M(Q, S, t)$ exists $= q$.

Proof   Suppose $mv(Q, S, t)$ exists $= q$. Therefore there is a sphere S_1 for which there are Q-measured-in-S-at-t worlds within S_1, and at all such worlds the mv of Q is q.[14] Now let M be the method of measuring Q in S at t at some such world w'. Obviously for all w' at which Q in S at t is measured by this same M and which are within S_1, the mv of Q in S at t at w' is q too (since they too are Q-measured-in-S-at-t-worlds). Hence there is a sphere, viz. S_1, for which within S_1 there are Q-measured-in-S-at-t-by-M-worlds and at all such worlds the mv for Q in S at t is q. Hence $mv_M(Q, S, t)$ exists $= q$. QED.

It is important however to note that the following is not of general validity:

(Prop)   If $mv_M(Q, S, t)$ exists $= q$ for all M then $mv(Q, S, t)$ exists $= q$.

We can construct a counter-example to (Prop) as follows. Suppose $mv_M(Q, S, t)$ exists and $= q$ for each of a countable infinity of different M—say M_1, M_2, ... This means that for each such M there is a sphere $S(M)$ for which there are Q-measured-by-M-worlds within $S(M)$ and at all such the mv of Q is q. (Here and from now on I abbreviate 'measured value of Q in S at t' to 'mv of Q'.) We also let $S(M)$ be the 'maximal' such spheres for each M, in that between $S(M)$ and any other sphere which is further from the actual world than $S(M)$ there is a Q-measured-by-M-world for which the mv of Q is *not* q. But suppose the sequence $S(M_1)$, $S(M_2)$, . . . is monotonic decreasing, i.e. $S(M_i) < S(M_{i-1})$ for all $i = 2, 3$. It follows that for each $S(M_i)$ there is a Q-measured-by-M_i-world which is outside $S(M_i)$ but which is within $S(M_{i-1})$, and at which the mv of Q is not q. Hence $mv(Q, S, t)$ does not exist since there is no sphere within which there are Q-measured-worlds all agreeing about the mv for Q. QED.

This counter-example to (Prop) depends critically on the infinite cardinality of the $\{M_i\}$. Indeed it is easy to prove that if $\{M_i\}$ is finite then (Prop) is valid; and this is my lemma 2:

Lemma 2   If $\{M_i\}$ is finite then (Prop) holds.

Proof   Suppose $mv_M(Q, S, t)$ exists $= q$ for all M, i.e. for each M there is a $S(M)$ for which there are Q-measured-by-M worlds within $S(M)$ at all of which the mv of Q is q. And let there be a finite number of such M. Now let $\bar{M}$ be the M such that $S(\bar{M})$ is the innermost of all the $S(M)$. (It is because $\{M\}$ is finite that we can assume such $\bar{M}$ exists.) But for any

---

[14]   I.e. for $mv(Q, S, t)$ to exist there must be a unique counterfactually construed mv, and hence some Q-measured in S at t worlds.

w which is a Q-measured-world within $S(\bar{M})$ there is an M for which w is a Q-measured-by-M-world. Moreover any such w is within $S(M)$ for any M (since $S(\bar{M})$ is *ex hypothesi* inside any $S(M)$). Hence at any such w the mv of Q is q. Hence there is a sphere, viz. $S(\bar{M})$, for which at any Q-measured-world within $S(\bar{M})$ (and there are such worlds) the mv of Q is q. That is mv(Q, S, t) exists $= q$. QED.

Now we are in a position to prove theorems 1 and 2, although, because of the failure of lemma 2 for the general case, we cannot strengthen Theorem 2 to:

(Pass Q)$_2$ implies (U-Pass Q).

Theorem 1   (Pass Q)$_2$ implies (U-Pass Q)$_1$.

Proof   Assume that (Pass Q)$_2$ and that $Q(S, t)$ exists with $Q(S, t) = q$ say. It follows that mv$_M(Q, S, t)$ exists $= q$ for each M. Hence at all near enough to the actual Q-measured-by-M-worlds (and there are such Q-measured-by-M-worlds—see footnote 14) the mv of Q is q. Moreover assume mv(Q, S, t) exists $= q'$ say. Then there is a sphere S_0 for which there are Q-measured-worlds w within S_0 and at each of which the mv of Q is q'. But each such Q-measured-world is, for some M, a Q-measured-by-M-world. Hence, there must be some M—say M^*—for which there are Q-measured-by-M^*-worlds within S_0. Moreover there are no Q-measured-by-M^*-worlds which are both closer to the actual than those and at which the mv of $Q \neq q'$ (otherwise mv(Q, S, t) would not be q'). Hence at all the Q-measured-by-M^* worlds within S_0—and there are some—the mv of Q must be q'. Hence mv$_{M^*}(Q, S, t) = q'$. But *ex hypothesi* mv$_{M^*}(Q, S, t) = q$. Hence $q = q'$. That is from the assumption that mv(Q, S, t) and $Q(S, t)$ exist and (Pass Q)$_2$ we have derived that $Q(S, t) = $ mv(Q, S, t), which means that from (Pass Q)$_2$ we can derive (U-Pass Q)$_1$. QED.

We can also prove:

Theorem 2   (Pass Q)$_2$ implies (U-Pass Q)$_2$ if there is only a finite number of measurements.

Proof   Assume $\{M_i\}$ finite, (Pass Q)$_2$ and that $Q(S, t)$ exists $= q$. Substituting that $Q(S, t) = q$ into (Pass Q)$_2$ gives that mv$_M(Q, S, t)$ exists $= q$ for all M. But now applying lemma 2 (which we can since $\{M_i\}$ is finite) it follows that mv(Q, S, t) exists $= q$. Thus in the case of $\{M_i\}$ finite, (Pass Q)$_2$ implies that if $Q(S, t)$ exists $= q$ then mv(Q, S, t) exists $= q$, i.e. (Pass Q)$_2$ implies (U-Pass Q)$_2$. QED.

And we can prove theorem 3 similarly, via lemma 1, as we can the following theorems for the (Pass Q)$'_2$ principles:

Theorem 4 (R-Pass Q)$'_2$ implies (U-Pass Q)$'_2$.

Theorem 5 (U-Pass Q)$'_2$ implies (Pass Q)$'_2$ if $\{M_i\}$ is finite.

Theorem 6 (U-Pass Q)$'_2$ implies (Pass Q)$'_2$.

CONCLUSION

I shall now summarize the results derived in this chapter. KS presents us with two initial options: to keep or not to keep (Det Q). Suppose we decide to keep (Det Q). Then we must admit QT to be incomplete, the most plausible way being to de-Ockhamize (see section 1). And if we also keep (R-Pass Q)$_1$ then we seem to be committed to a revision of QMT which severely restricts the range of measurements for degenerate physical quantities (see end of section 2).

We then have to make a further choice whether or not to assume the existence of all the mv(Q, S, t) (and this was discussed in section 3). If we do not take them all to exist, for some S and t, then we must give up (U-Pass Q)$_2$, since there are then some possessed values not matched by mvs. On the other hand if we take all the mv(Q, S, t) to exist, and still persist with standard QMT, then we find that we must contextualize our measured values.

Now consider the second of the two options which we faced initially—that of rejecting (Det Q). This removes the need to de-Ockhamize, and so, for this option, we ought not de-Ockhamize (since the onus of justification surely lies with the de-Ockhamizer). Again we must choose whether or not to accept that all the mv(Q, S, t) exist. If we do not take them all to exist (for some S, t) as Bohr does, then we at least have the possibility of rescuing all of the (Pass Q) principles (although Bohr does not take advantage of this possibility, being committed to rejecting both (Pass Q)$'_2$ and (U-Pass Q)$'_2$). On the other hand if we assume all the mvs exist then given standard QMT and (Func)$_0$ we must contextualize measured values, i.e. allow mvs to depend on the method of measurement; although, since on this sub-option we do not assume (Det Q), we do not need to go as far as we did above and admit that some ideal measurements are inaccurate. On this sub-option, however, (Pass Q)$'_2$ principles must be given up, since (Det Q) fails but the mvs all exist.

On the basis just of the arguments considered here, I suggest that it is the Bohr option of denying (Det Q) and the existence of all the mvs which is to be preferred. This is because the Bohr option is the only one of the options considered which retains all of standard QT (and QMT) without the need to de-Ockhamize physical quantities or contextualize measured values, and thus has an advantage in simplicity. Moreover its denial of the simultaneous existence of mvs (counterfactually construed) for strongly incommensurable physical quantities can, as we have seen, be justified independently by setting up a possible world analysis of the claim that measurements in QT are intrinsically statistical (or by simply taking the measurements as indeterministic). In particular there is no need to appeal to verificationist principles to justify Bohr's rejection of the coexistence of the mvs of incommensurables.

Of course Bohr's option does involve a departure from what I argued (in the previous chapter) are somewhat misleadingly called 'realist principles', viz. (Det Q) and (Pass Q), but as we have seen—and this is one of the central results of this chapter—some such departure is inevitable if we wish to retain the framework of standard QT. Bohr's option involves no departure from realism in any other respects however; and in particular does not involve 'anti-realism' in the sense of verificationism.

In the two following chapters I shall show that these conclusions are reinforced by arguments by Stapp and Eberhard, and by Heywood and Redhead. The Stapp and Eberhard arguments show that even de-Ockhamization does not save the realist, although this result relies on building in a locality condition to realism. And the Heywood–Redhead argument shows that, even without a locality clause, realism cannot be reconciled with standard QT, even one which has been contextualized and de-Ockhamized.

The question of whether Bohr's option is still the best option, when Schrödinger's cat paradox or the EPR paradox are taken into account, is however another question. It may well be that Bohr's interpretation ought to be modified in order to cope with the paradoxes (e.g. by incorporating the density operator criterion (Bohr)′ for physical quantities possessing values), even though, as I have argued, this is at the cost of reintroducing the complexities of de-Ockhamization or contextualization. It is this tentative conclusion towards which the arguments of this book have been heading.

8

Stapp and Eberhard

Aspect's experiment; Stapp's proof; introducing counterfactuals; freedom of measurement; locality principles; ropey worlds objection; de-Ockhamization.

INTRODUCTION

How do we explain the results of measurement in QT? Two ways suggest themselves within a broadly realist framework. We could assume determinate possessed values for the q-quantities, which are then reflected by (an implicitly idealized) measurement process. In other words we have the traditional 'hard' realist position consisting of the combination of (Det Q) and (Pass Q). In addition it is usually assumed that these possessed values do not just come into existence when the measurements are made (*contra* Feyerabend's explication of Bohr), but are independent features of an otherwise hidden level of reality which are displayed whenever measurements happen to take place.

Or we could adopt a 'softer' realist position in which the values appearing on measurement are merely taken as manifestations of underlying permanent dispositions to produce certain values on measurement. (Heisenberg's ontology of *potentia* is an example of such a position.)[1] The latter view is still realist in the broad sense that a permanent level of reality is assumed to exist 'behind the veil of perception', one which acts as a seat for the dispositions. But the latter view is 'softer' than the former position in the sense that it does not assume (Det Q) or (Pass Q). The principle which is characteristic of the latter soft realist position is the (MV) of the previous chapter:

[1] See Heisenberg (1971).

(MV)   For any q-system S and time t during the lifetime of S and any
        q-quantity Q for S there is a unique 'measured value' for Q in S
        at t, viz. the value which Q would be measured to have were it
        measured in S at t.

In Chapter 7 I argued against (MV) on what were essentially
metaphysical grounds, and in particular from within a particular
interpretation of the intrinsic probabilities of QT. In this chapter I
shall reinforce this earlier conclusion by arguing against (MV) from
within QT itself, assuming only a fairly uncontroversial metaphysical
framework. In particular, in section 1 I shall present a version of a
well-known argument by Stapp (1979) and (in another version) by
Eberhard (1977), based on a theorem proved by Bell, to the effect that
the usual laws of QT are inconsistent with a combination of the
following three metaphysical principles: (MV), a principle of 'local
causation' (roughly that there is no instantaneous action at distance),
and what I shall call 'a principle of freedom of measurement'—'(FM)'
for short—which denies the existence of certain sorts of causal
restrictions on what can be measured. In other words it will be shown
that we cannot reconcile QT with soft realism given the two other
classical metaphysical principles (FM) and locality. The proof works
in effect by showing that QT requires not just the algebraic, i.e.
ontological, contextualization discussed in Chapter 7. It also requires
what Shimony (1984) (appropriating a term proposed by Heywood
and Redhead (1985)) calls 'environmental contextualization': that the
mv of Q for S at t may depend on what else, other than Q, is measured
at a distance from S at t. This extra dimension to contextualization is
then shown to conflict with locality requirements.

This same argument can then be trivially extended to show that
hard realism is inconsistent with a combination of QT, locality, and
(FM) as follows. If we assume the realist principles of passivity of
measurement and the existence of determinate possessed values for
q-quantities then we are clearly committed to the existence of
measured values, viz. measured values to match the *ex hypothesi*
existing possessed values. But the existence of these measured values
is inconsistent with a combination of QT, locality, and (FM), and
hence so are these hard realist principles.

Some realists may consider (Pass Q) to be too strong an assumption to be part of a classical realist metaphysics in the first place and
hence will find no difficulty in giving up (Pass Q) in response to the

previous proof. (A similar move, it will be remembered, can be made in response to the results of the previous chapter.) But this concession is of little avail, because the previous proof will also work if we assume a weaker, and more realistic, version of (Pass Q), viz. that possessed values can always be measured to some small but finite degree of accuracy. In other words, we do not need to assume the existence of perfect measurements, but only the existence of measurements which are 'good enough'.

In short, in this chapter I shall be concerned to show how certain realist principles (both hard and soft), together with a locality principle, conflict with QT given (FM). Moreover I shall take this conflict, together with the broad support for QT and the taken for granted status of (FM), to argue for the rejection of local realism. However, I shall play down the significance of this as an argument against realism, as indicated in Chapter 6. More to the point, following the argument of Chapter 7, I shall argue that the relevant rejection of realism can be achieved merely by introducing a form of indeterminism strong enough to imply the non-existence of certain counterfactually definite measured values; in other words we may merely have a failure of what Stapp calls 'counterfactual definiteness'.

In this chapter I shall also take issue with the way in which the argument for the failure of local realism is presented in the literature, e.g. by d'Espagnat (1979). D'Espagnat considers local realism to be *experimentally* refuted by the evidence from experiments such as Aspect's (1982). These experimental 'refutations' rely on the fact that local realism (following along the lines of the proof I shall give below) can be shown to imply an inequality, viz. Bell's inequality, which not only contradicts QT—thus making the point I made above, viz. that local realism and QT conflict—but also is open to direct experimental testing. This fact, that the contradiction between QT and local realism is manifested at the observational level, then allows us to test both QT and local realism as part of the same testing procedure, viz. testing Bell's inequality. We have in short precisely the circumstances for a classic Baconian 'crucial experiment'. And, d'Espagnat claims, the results of this crucial experiment refute local realism and confirm QT (by refuting Bell's inequality).

My quarrel here is not with d'Espagnat's dismissal of local realism, but with his claim that it is the results of Aspect's experiment (and several other such experiments) which *by themselves* refute local realism. I claim on the contrary that it is the formal conflict between

QT and local realism, together with the support for QT at large, which allows us to refute local realism, rather than the results of any crucial experiments to test Bell's inequality. In saying this I am not denying the logical point that Bell's inequality contradicts QT. Rather I am making the familiar epistemological point that the results of experiments can always be reinterpreted (as consistent with Bell's inequality say, rather than inconsistent with it); and moreover that this possibility for reinterpretation is typically mobilized in order to achieve coherence with the theoretical context into which the results are inserted. Thus, as Lakatos argues (1970) p. 91ff., the way we interpret the results of our experiments is rarely (perhaps never) so as to contradict some accepted theory.[2] It is only when those experimental results can be relocated (and reinterpreted) in the context of some viable rival theory that such a contradiction can arise. Thus it is the existence of a viable rival to local realism, viz. QT, within which the results of Aspect's experiment can be located, as much as the experimental results themselves, which allows us to see the results of Aspect's experiment as refuting local realism.[3] And of course this conflict with QT, given the wide independent support for QT, is already enough to count against local realism even before we come to consider the results of any crucial experiments. This is not to deny that the results of Aspect's experiment are valuable extra confirmation for QT and against local realism. But it is to shift the emphasis away from such confirmation as the sole, or indeed most important, reason for rejecting local realism.

One point which may strike the reader here is my preference for the Stapp and Eberhard modalized versions of what is essentially the no-local-hidden-variables argument put forward by Bell, e.g. see Bell (1981). Bell's original argument used statistical terminology, and this may actually seem an advantage in that it thereby avoids the contentious use of counterfactuals. But, as I argued at the end of Chapter 3, Bell's argument presupposes the Reichenbach principle of prior common cause, which is precisely what must be questioned if we accept that QT is indeterministic. Therefore, to avoid begging the question of determinism, I shall use the Stapp and Eberhard versions

---

[2] There is a difficulty here with the category 'theory': if it includes 'low-level' observational generalization then Lakatos's claim becomes implausible. But for higher-level theories, such as local realism and QT, Lakatos's claim is more plausible.

[3] This phrase 'the results themselves' as it appears in the last sentence should not be taken to imply that there is some context-free interpretation of the results.

of Bell's argument. (This is not of course to say that Bell's original argument is not useful in its own right. Clearly it is as a means of ruling out a certain broad class of stochastic hidden variables theories. But that is not my concern here.)

Note that in this chapter I shall only have recourse to mention probabilities of the form $P(Q, q, f)$. On the scheme I have been using here the latter probability should be interpreted as shorthand for '$P(Q, q, S, t)$' in the special case where $f(S, t) = f$. But more generally we can take it as the 'probability for Q-measurements on q-systems which are in the state f of registering the value q'. This interpretation bypasses the contentious issue of interpreting $P(Q, q, S, t)$, and is broad enough to accommodate orthodox interpretations of QT as well as a hidden variables approach (on which $P(Q, q, S, t)$ is construed as an epistemic probability reducible to 1 or 0 once we relieve our ignorance about S at t). Also note that in this chapter I shall mention probabilities of the form $P((Q, Q'), (q_i, q'_j), F)$, which is the joint probability of registering the values q_i and q'_j respectively for simultaneous measurements of Q and Q' on different parts of some joint q-systems in the state F. (Within QT we can take this as $P(Q'', q_{ij}, F)$, where Q'' is a single physical quantity which has value q_{ij} with certainty for some joint system at t iff both Q has value q_i with certainty for one part of the system at t and Q' has value q'_j with certainty for the other part of the system at t.) Finally, note that the probabilities will sometimes appear with a superscript—thus '$P^1(Q, q, F)$'—in order to indicate that the probability is being considered at a particular world—say the possible world w_1.

1. THE STAPP AND EBERHARD ARGUMENT

In this section I shall derive a contradiction between QT, locality, (FM), and (MV). Consider a pair of electrons in a pure spin 0 'singleton' state F at time t. We shall distinguish the members of the pair by referring to one of them as 'L' (on the left) and the other as 'R' (on the right). Measurements of the spin-components of the L and R electrons are envisaged as being made by apparatuses located in the vicinity of L and R respectively. Technical details of what sort of q-quantities the spin-components of an electron are will not concern us here, except to say that every electron has a q-quantity called 'the spin-component of the electron along the direction $\mathbf{a}$' and there is such a q-quantity for every possible direction **a**, each such q-quantity having possible values $+1$ or -1 and no other possible values. For convenience I shall let 'a' denote not just the direction **a** but also the spin-component of an electron along the direction **a**.

The spatial separation of the L and R electrons is taken as so large that, unless we allow a 'non-local' (faster-than-light) transmission of effects, the outcome of a spin-component measurement on L at t cannot be affected by states of affairs at t near R, in particular it cannot be affected by which spin-component is measured on R at t.

Now consider a collection C of a number N of such electron-pairs, all in the singlet state F at t, the L-member of any one such pair being spatially well separated from the R-member; and also suppose that any of the pairs is spatially well separated from all other pairs at t. Suppose that next to each L-electron is a spin-measuring apparatus, measuring spin in either the **a** or **a**′ direction at t, and next to each R-electron is a spin-measuring apparatus measuring spin in either the **b** or **b**′ direction at t. Both **a** and **a**′ cannot be measured at t on the same electron since **a** and **a**′ are incommensurable, and similarly for **b** and **b**′. But assume (for Bohr *per impossibile*) that there are counterfactually construed measured values for both **a** and **a**′ at t for each L-electron in C, and for both **b** and **b**′ at t for each R-electron. (This assumption is part of what in Chapter 7 we called (MV), one of the assumptions for which I am here providing a joint *reductio*.) Let 'a_n' and 'a'_n' be rigid designators for the numerical values taken by the counterfactual measured values for **a**, **a**′ respectively for L_n, the L-electron in the nth pair; and let 'b_n' and 'b'_n' be the corresponding rigid designators for the numerical values taken by the measured values for **b** and **b**′ respectively for R_n, the R-electron in the nth pair. Each of these values will be either $+1$ or -1, the possible values for the spin-components. For convenience I assume here that all the counterfactual measured values are strictly counterfactual, i.e. no measurements in fact take place. Note, however, that I am not interpreting 'a_n' strongly as the value which **a** would be measured to have *whatever else* is measured. In particular (MV) does not build in the assumption that all ways of measuring **a** would have the same result. It merely assumes, in Lewis's terms, that all possible ways of measuring **a** on L_n at t which are near enough to the actual world have the same result, viz. a_n.

Now following the proofs by Stapp (1979) and Eberhard (1977)[4] we

---

[4] For a general summary of such proofs see Clauser and Shimony (1978), including a generalization to imperfect measurements.

define

$$G_n = a_n b_n + a_n b'_n + a'_n b_n - a'_n b'_n$$

$$= a_n(b_n + b'_n) + a'_n(b_n - b'_n).$$

But either b_n has the same sign as b'_n or the opposite sign. In the first case $|G_n| = 2|a_n| = 2$. And in the second case $|G_n| = 2|a'_n| = 2$. Thus $|G_n| = 2$.

Now consider

$$\sum_{n=1}^{N} |G_n|/N = \Sigma|a_n b_n + a_n b'_n + a'_n b_n - a'_n b'_n|/N \geq |\Sigma(a_n b_n$$

$$+ a_n b'_n + a'_n b_n - a'_n b'_n)/N|$$

$$= |\Sigma a_n b_n/N + \Sigma a_n b'_n/N + \Sigma a'_n b_n/N - \Sigma a'_n b'_n/N|.$$

If we define $E(\mathbf{a}, \mathbf{b}) = \Sigma a_n b_n/N$, and similarly define $E(\mathbf{a}, \mathbf{b}')$, $E(\mathbf{a}', \mathbf{b})$, $E(\mathbf{a}', \mathbf{b}')$ then we see that the right-hand side of this last equality is

$$| E(\mathbf{a}, \mathbf{b}) + E(\mathbf{a}, \mathbf{b}') + E(\mathbf{a}', \mathbf{b}) - E(\mathbf{a}', \mathbf{b}') |,$$

whereas clearly, since $|G_n| = 2$, the left-hand side of the first equality is 2. Hence

(i) $| E(\mathbf{a}, \mathbf{b}) + E(\mathbf{a}, \mathbf{b}') + E(\mathbf{a}', \mathbf{b}) - E(\mathbf{a}', \mathbf{b}') | \leq 2.$

(This last inequality is one of a series of inequalities which have been developed from Bell's work—see Clauser and Shimony (1978).)

The task now is to show, with the help of (FM) and a locality assumption, that there is an at least approximate identity between the various $E(\mathbf{a}, \mathbf{b})$, $E(\mathbf{a}, \mathbf{b}')$, ... and certain quantities evaluated in QT. Having done this (i) can be tested against QT, and will be found to be in contradiction with it. It is in this way that we can conclude that (MV), which was used to derive (i), together with (FM) and the locality assumption, contradict QT.

I shall first of all carry out this proof in a rather sketchy fashion in order to provide an overview of the whole proof. Were $\mathbf{a}$ to have been measured on L_n at t then the mv of $\mathbf{a}$ would have been a_n, by definition of 'a_n'. But now ask the question: were not only $\mathbf{a}$ to have been measured on L_n at t but also $\mathbf{b}$ measured on R_n at t what would the mv of $\mathbf{a}$ on L_n at t have been? Would it still have been a_n? Yes, so goes the argument, because what goes on in R_n at t is spatially well separated from L_n at t, and hence can have no effect on the outcome of a measurement on L_n at t, at least not if we restrict ourselves to 'local' causal processes (i.e. if we disallow the faster-than-light transmission of causes, i.e. disallow instantaneous action at a distance on the special relativistic construal of 'instantaneous'). A similar conclusion can be derived for b_n.

The step just made is the crucial step in the proof. It rules out what in the introduction we called 'environmental contextualization', i.e. the dependence of mvs on the context of measurement, in so far as that context includes what else is measured. Moreover we see that this contextualization has been ruled out by appeal to a principle of locality. In short there is an inconsistency implicitly being asserted here between locality and environmental contextuality. This inconsistency will be argued for later.

We now ask the further question: what is the value which **a** would be measured to have on L_n at t were **a**, **b** measured on all the L_n, R_n at t for all n? Again the answer is 'a_n'; and it is again a locality condition which tells us this, i.e. the large spatial separation of the various pairs from each other at t means that what is measured on the other pairs at t does not affect the outcome of measurements on the nth pair at t. And this is so for all n, and similarly for **b** instead of **a**. Hence rf$(($**a**, **b**$), (a, b))$, the relative frequency with which (a_n, b_n) is equal to (a, b) over the various n, is equal to the rf with which **a**, **b** would be measured to have values (a, b) were **a**, **b** measured on *all* the pairs in C.

But surely it is plausible to take the latter rf as $P(($**a**, **b**$), (a, b), F)$ to a good approximation for large enough N. (Implicit here is something like the law of large numbers; although, as we shall see, this implication is not all that straightforward because we are here considering a counterfactual relative frequency.) And hence, to a good approximation for N large enough, we can equate rf$(($**a**, **b**$), (a, b))$ with $P(($**a**, **b**$), (a, b), F)$.

But it is easy to prove that

$$E(\mathbf{a}, \mathbf{b}) = \sum_{a,b} (\text{rf}((\mathbf{a}, \mathbf{b}), (a, b))) \times (ab).^5$$

[5] $E(\mathbf{a}, \mathbf{b}) = \sum_n a_n b_n / N.$

Now collect together all those terms of form '$a_n b_n$' for which $a_n = a$ and $b_n = b$. Let there be $N(a, b)$ of them. Thus

$$E(\mathbf{a}, \mathbf{b}) = \sum_{a,b} \sum_{n:\, a_n = a,\, b_n = b} (a_n b_n)/N$$
$$= \sum_{a,\, b} N(a, b) \times (ab)/N$$
$$= \sum_{a,\, b} \text{rf}((\mathbf{a}, \mathbf{b}), (a, b)) \times (ab).$$

And hence, if we define '$\langle \mathbf{a} \times \mathbf{b} \rangle_F$' as shorthand for:

$$\sum_{a,b} (P((\mathbf{a}, \mathbf{b}), (a, b), F)) \times (ab),$$

then we see that to a good approximation $E(\mathbf{a}, \mathbf{b}) = \langle a \times b \rangle_F$. And hence to a good approximation the left-hand side of (i) is:

(i)' $|\langle \mathbf{a} \times \mathbf{b} \rangle_F + \langle \mathbf{a} \times \mathbf{b'} \rangle_F + \langle \mathbf{a'} \times \mathbf{b} \rangle_F - \langle \mathbf{a'} \times \mathbf{b'} \rangle_F|$.

But this $\langle \mathbf{a} \times \mathbf{b} \rangle_F$—the 'expectation value of $\mathbf{a} \times \mathbf{b}$ for state F'—is a quantity the value of which QT tells us (by telling us the various possible values (ab) of $\mathbf{a} \times \mathbf{b}$ as well as the corresponding probabilities $P((\mathbf{a}, \mathbf{b}), (a, b), F)$). In particular it tells us that (i)' > 2 by some fixed amount which is independent of N for given $\mathbf{a}, \mathbf{b}, \mathbf{a'}, \mathbf{b'}$. Indeed (i)' may be as much as $5/2$ for certain $\mathbf{a}, \mathbf{b}, \mathbf{a'}, \mathbf{b'}$. Hence, as required, we have derived a contradiction with (i).

I shall now fill out the previous proof sketch by introducing a complication to get around the gap when we equated the rf$((\mathbf{a}, \mathbf{b}), (a, b))$ with $P((\mathbf{a}, \mathbf{b}), (a, b), F)$. For convenience I shall work through the full proof initially in terms of a Lewis possible-world semantics. First of all we assume that there is a non-empty set of possible worlds $S(\mathbf{a}, \mathbf{b})$ for which at each w in $S(\mathbf{a}, \mathbf{b})$ there is a counterpart for each electron pair in the real-world collection C described above, but these counterparts differ from the real-world members of C in the respect that $(\mathbf{a}, \mathbf{b})$ is measured on all of the electron-pairs in C at t at w. Moreover we assume that QT holds at all the near enough to the actual world members of $S(\mathbf{a}, \mathbf{b})$. This last assumption—call it '$(FM)_1$'—can be seen as an aspect of 'Freedom of Measurement' because it follows from the principle that QT would still hold (as it does in the actual world) even were we to measure $(\mathbf{a}, \mathbf{b})$ on all the members of C at t. In other words we are free to measure $(\mathbf{a}, \mathbf{b})$ on all members of C without prejudicing the laws of QT.

We also assume—call it '$(FM)_2$'—that all near enough to the actual world members of $S(\mathbf{a}, \mathbf{b})$ are worlds at which all the pairs in C are in the state F at t. Again this is an aspect of (FM) since it says that even were we to measure $(\mathbf{a}, \mathbf{b})$ on all members of C at t the states of the measured systems at t would still stay the same, viz. F; i.e. the fact of measurement alone would leave untouched the states of the measured system leading up to the measurement. (Although the states during and after the measurement may of course be affected by the measurement.)

From $(FM)_1$ and $(FM)_2$ it then trivially follows that

(A)   For any w in $S(\mathbf{a}, \mathbf{b})$ there is a w_1 no further from the actual world than w at which not only is $\mathbf{a}, \mathbf{b}$ measured on all the pairs in C at t but also all the pairs are in state F at t and the laws of QT hold.

Now let C_1 be the counterpart of C at such a w_1, i.e. C_1 is made up of counterparts at w_1 of the very pairs which make up C at the actual world. For the collection C_1, unlike C, it is uncontroversial that the measured value a_n^1 of $\mathbf{a}$ for L_n exists for all n (because $\mathbf{a}$ is in fact measured on all L_n for C_1); and similarly it is uncontroversial that b_n^1, the measured value of $\mathbf{b}$ for R_n, at w_1 exists for all n. In particular $E^1(\mathbf{a}, \mathbf{b})$ exists, where we define

$$E^1(\mathbf{a}, \mathbf{b}) = \sum a_n^1 b_n^1 / N.$$

Hence (as proved in footnote 5)

$$E^1(\mathbf{a}, \mathbf{b}) = \sum_{a, b} \mathrm{rf}^1((\mathbf{a}, \mathbf{b}), (a, b)) \times (ab).$$

But the set of $\mathbf{a}, \mathbf{b}$ measurements on the electron pairs in C_1 constitutes a Bernoullian set in that (it is easily shown from within QT)[6] the outcomes of the measurements are statistically independent, essentially because each pair is in a pure state F, *and* the probability of a particular outcome—say a, b as measured values—is constant across the various members of the set, i.e. it is just $P^1((\mathbf{a}, \mathbf{b}), (a, b), F)$. Hence the law of large numbers tells us that

$$\mathrm{rf}^1((\mathbf{a}, \mathbf{b}), (a, b)) \approx P^1((\mathbf{a}, \mathbf{b}), (a, b), F),$$

where the approximation here is statistical in nature; in particular the probability of there being any given difference between the rf^1 and the corresponding P^1 can be made as small as we like by letting N be large enough. Indeed the probability can be made 0 by letting N go to infinity. (It is the Strong law of large numbers which tells us this—see Feller (1950). Note that it is a non-trivial fact about the systems considered here that we can actually go to the infinite limit. Essentially it is because each system has a finite dimensional Hilbert space and because a countably infinite product of finite dimensional Hilbert spaces is still separable, and hence within the purview of QT.)

---

[6] See Appendix 11.

Hence we can see that $E^1(\mathbf{a}, \mathbf{b})$, for large N, is a 'good estimator' of $\langle \mathbf{a} \times \mathbf{b} \rangle_F^1$, the expectation value of $\mathbf{a} \times \mathbf{b}$ for electron pairs in state F at w_1, which is defined as:

$$\sum_{a,b} P^1((\mathbf{a}, \mathbf{b}), (a, b), F) \times (ab).$$

Since the laws of QT hold at w_1 the $P^1((\mathbf{a}, \mathbf{b}), (a, b), F)$ can be identified as the $P((\mathbf{a}, \mathbf{b}), (a, b), F)$ determined by QT; and *a fortiori* $\langle \mathbf{a} \times \mathbf{b} \rangle_F^1$ can be identified with the $\langle \mathbf{a} \times \mathbf{b} \rangle_F$ of QT.

The next step in the full proof is that of proving that, for a suitable choice of w and hence w_1 in $S(\mathbf{a}, \mathbf{b})$, $E(\mathbf{a}, \mathbf{b}) = E^1(\mathbf{a}, \mathbf{b})$; and we do so with the help of (what I shall later show to be) a locality assumption:

(Strong Loc)　Were $(\mathbf{a}, \mathbf{b})$ measured on *all* the electron-pairs at t (with $\mathbf{a}$ measured on the L-electrons, $\mathbf{b}$ on the R-electrons) then the mv of $\mathbf{a}$ for L_n at t would be a_n (where a_n it will be remembered is a rigid designator for the numerical value which $\mathbf{a}$ would be measured to have were it measured on L_n at t). And the same holds if $\mathbf{a}$ and L_n are interchanged with $\mathbf{b}$ and R_n, and for all n.

On a Lewis analysis (Strong Loc) implies that

For any n, for any w which is both a member of $S(\mathbf{a}, \mathbf{b})$ and near enough to the actual world, the mv of $\mathbf{a}$ for L_n at t at w is a_n and the mv of $\mathbf{b}$ for R_n at t at w is b_n.

(Note here the importance of 'a_n' being a rigid designator, i.e. not being world-dependent.)

Hence (and since $S(\mathbf{a}, \mathbf{b})$ is non-empty)

(B)　For any n there is a w_2 in $S(\mathbf{a}, \mathbf{b})$ for which, for any w in $S(\mathbf{a}, \mathbf{b})$ which is no further from the actual than w_2, the mv of $\mathbf{a}$ for L_n at t at w is a_n and the mv of $\mathbf{b}$ for R_n at t at w is b_n.

Now if we retrospectively substitute the w_2 in (B) for the w in (A) above then we see that for the corresponding w_1 introduced in (A) the mvs of $\mathbf{a}, \mathbf{b}$ at t for L_n, R_n respectively at w_1 are a_n, b_n respectively, and this is so for any n. Hence $E^1(\mathbf{a}, \mathbf{b}) = E(\mathbf{a}, \mathbf{b})$, and so $E(\mathbf{a}, \mathbf{b})$ is also a good estimator of $\langle \mathbf{a} \times \mathbf{b} \rangle_F$—as good as (we have shown) $E^1(\mathbf{a}, \mathbf{b})$ is.

We can then repeat this same derivation three more times but with '$(\mathbf{a}, \mathbf{b}')$', '$(\mathbf{a}', \mathbf{b})$', and '$(\mathbf{a}', \mathbf{b}')$' respectively instead of '$(\mathbf{a}, \mathbf{b})$'; and hence conclude that $E(\mathbf{a}, \mathbf{b}')$, $E(\mathbf{a}', \mathbf{b})$, $E(\mathbf{a}', \mathbf{b}')$ are good estimators of $\langle \mathbf{a}$

$\times \mathbf{b}'\rangle_F$, $\langle \mathbf{a}' \times \mathbf{b} \rangle_F$, and $\langle \mathbf{a}' \times \mathbf{b}' \rangle_F$ respectively, where for any d for N large enough the probability that the estimates are out by more than d becomes as low as one likes. Thus for N large enough it is, it seems, unreasonable to refuse to identify $\langle \mathbf{a} \times \mathbf{b} \rangle_F$ with $E(\mathbf{a}, \mathbf{b})$. More generally $|E(\mathbf{a}, \mathbf{b}) + E(\mathbf{a}, \mathbf{b}') + E(\mathbf{a}', \mathbf{b}) - E(\mathbf{a}', \mathbf{b}')|$ is as good an estimate as one likes of $|\langle \mathbf{a} \times \mathbf{b} \rangle_F + \langle \mathbf{a} \times \mathbf{b}' \rangle_F + \langle \mathbf{a}' \times \mathbf{b} \rangle_F - \langle \mathbf{a}' \times \mathbf{b}' \rangle_F|$ for N large enough. And this in turn means (given (i)) that for N large enough it is reasonable to assert:

(iv) $\quad |\langle \mathbf{a} \times \mathbf{b} \rangle_F + \langle \mathbf{a} \times \mathbf{b}' \rangle_F + \langle \mathbf{a}' \times \mathbf{b} \rangle_F - \langle \mathbf{a}' \times \mathbf{b}' \rangle_F| > 2.$

But QT implies:

(v) $\quad |\langle \mathbf{a} \times \mathbf{b} \rangle_F + \langle \mathbf{a} \times \mathbf{b}' \rangle_F + \langle \mathbf{a}' \times \mathbf{b} \rangle_F - \langle \mathbf{a}' \times \mathbf{b}' \rangle_F| \leq 2.$

Hence it is unreasonable to assert all of QT, (MV), (FM), and (Strong Loc). In short we have not so much demonstrated a formal contradiction between (QT), (MV), (FM), and (Strong Loc) as a 'pragmatic contradiction', viz. that it is unreasonable to assert all of them together. QED.

Note that this argument will still hold even if we consider only approximately ideal measurements, for which the Born statistical interpretation only holds approximately, subject only to the restriction that the relevant approximation is good enough not to affect the contradiction between (iv) and (v). Since the left-hand side in (iv) can be made as large 5/2 this last restriction can be met in practice.[7]

The previous argument can also be seen to imply that realism *qua* the combination of (U-Pass Q)$_2$ (that if possessed values exist then so do measured values, and they equal the possessed values) and (Det Q) is not (pragmatically) consistent with QT, (FM), and (Strong Loc). Suppose $\mathbf{a}, \mathbf{a}', \mathbf{b}, \mathbf{b}'$ all have determinate values at t for the electrons in C (following (Det Q)). If we then assume (U-Pass Q)$_2$ we derive the existence of the mvs for $\mathbf{a}_n, \mathbf{a}'_n, \mathbf{b}_n, \mathbf{b}'_n$, which, in order to preserve (Strong Loc) and (FM), we have just rejected. Hence the realist must reject at least (U-Pass Q)$_2$ if he is to preserve (Det Q), (FM), and (Strong Loc).

The previous argument can also be seen to imply that even on the modified interpretation of QT (with (Bohr)' but without full (Det Q)) some aspect of passivity must be given up. If electron pair $S_1 + S_2$ at t is in the singlet state F then it is easy to show that $\hat{W}(S_1, t)$, the density

---

[7] See Clauser and Shimony's discussion of measurement techniques in their (1978).

operator for S_1 at t, is the identity operator on $\mathcal{H}(S_1)$; and as such, by (Bohr)', each q-quantity for S_1, including both **a** and **b**, has a determinate value at t; and similarly for **a**', **b**', the q-quantities for S_2.[8] The argument of the previous section can then be repeated if we merely assume a strong enough passivity principle to guarantee the existence of the relevant mvs. Thus we can show that the interpretation of QT which I have been endorsing here (which includes (Bohr)') is also in contradiction with passivity, if we retain (Strong Loc) and (FM). Moreover this argument will still hold even if the passivity principle is weakened to an approximate identity between possessed and measured values (with the qualification made above).

Note finally that the previous argument can be carried out without using the Lewis possible world apparatus. I shall here only sketch how this is to be done. Abbreviate '**a**, **b** are measured on L_n, R_n at t for all n' by 'M_{ab}'. Abbreviate '**a** is measured on L_n at t to have value a_n' by 'a_n'. Abbreviate 'L_n, R_n at t are in the state F for all n' by 'F'. Then (Strong Loc) says, for any n,

$$M_{ab} \to a_n \quad \text{and} \quad M_{ab} \to b_n.$$

Hence (by the inference rule: $p \to q, p \to r \therefore p \to q.r$).

(i) $\quad M_{ab} \to (a_n.b_n \text{ for all } n)$.

Now assume, as aspects of (FM) (freedom of measurement), that whether QT holds is independent (causally) of what is measured on all the pairs in C_N at t *and* that the states of the various pairs in C_N at t are also independent of what is measured on those pairs at t. On a Lewis analysis of causal independence[9] this means that

(ii) $\quad M_{ab} \to \text{QT and} \sim M_{ab} \to \text{QT}.$

(iii) $\quad M_{ab} \to F \text{ and} \sim M_{ab} \to F.$

Now bringing together (i), (ii), and (iii) gives:

(iv) $\quad M_{ab} \to ((a_n.b_n \text{ for all } n).\text{QT}.F).$

From (iv) and the extra assumption $\text{Poss}(M_{ab})$ (where 'Poss' is the

---

[8] The value must be *determinate* of course if or-elimination is to hold for the disjunction in 'a_n has value $+1$ or -1'. A quick look at the above argument will show that we do indeed need to make precisely that assumption.

[9] See Lewis (1975). We take causal independence as the obvious contrary of his causal dependence.

operator 'It is possible that') it follows that

Poss$((a_n \cdot b_n$ for all $n)$. QT. $F)$.

(By the inference rule: $p \rightarrow q$, Poss $p \therefore$ Poss q.)

But '$a_n \cdot b_n$ for all n' clearly implies that, for arbitrary $a, b,$ rf$(\mathbf{a}, \mathbf{b}$ meas $a, b)$, the relative frequency with which $\mathbf{a}, \mathbf{b}$ are measured to have values a, b over the N pairs at $t, =$ rf$((\mathbf{a}, \mathbf{b}), (a, b))$, the relative frequency with which a_n, b_n are equal to a, b over the N pairs. Hence

Poss$(($rf$(\mathbf{a}, \mathbf{b}$ meas $a, b) =$ rf$((\mathbf{a}, \mathbf{b}), (a, b)) \cdot (a_n \cdot b_n$ for all $n)$. QT. $F)$.

But '$(a_n \cdot b_n$ for all $n)$. QT. F' implies that rf$(\mathbf{a}, \mathbf{b}$ meas $a, b) \approx |(F, f_{\mathbf{a}, a} \times f_{\mathbf{b}, b}|^2$ (see above). Hence

Poss$($rf$((\mathbf{a}, \mathbf{b}), (a, b)) \approx |(F, f_{\mathbf{a}, a} \times f_{\mathbf{b}, b}|^2)$.

But the expressions on the left-hand and right-hand side of $\approx$ here are rigid designators of numbers; and hence if they are possibly equal (equal in some possible world) then they are equal *simpliciter* (indeed equal in all possible worlds). Thus we derive the identity:

rf$((\mathbf{a}, \mathbf{b}), (a, b)) \approx |(F, f_{\mathbf{a}, a} \times f_{\mathbf{b}, b}|^2$;

and hence of course

$E(\mathbf{a}, \mathbf{b}) \approx \langle \mathbf{a} \times \mathbf{b} \rangle_F$.

Similarly we derive the other identities, and the proof can then be completed as above.

2. RESPONSES TO STAPP AND EBERHARD

One way to accommodate the conclusion of the previous section is to claim that the principle (Strong Loc) is not purely a locality principle, but involves something more which can be rejected without introducing non-localities. And indeed (Strong Loc) may fail simply by the non-existence of mv$_{\mathbf{ab}}(\mathbf{a}_n)$, the counterfactually construed mv of $\mathbf{a}$ on L_n at t were $(\mathbf{a}, \mathbf{b})$ measured on all the members of C; and this non-existence is clearly not itself strong enough to guarantee a non-local effect. ((Strong Loc) could also of course be taken to fail just because there is no mv$(\mathbf{a}, L_n, t)$, i.e. no mv$(\mathbf{a}_n)$, *or* no mv$(\mathbf{b}, R_n, t)$, i.e. no mv$(\mathbf{b}_n)$, for some n; but this would contradict the assumption (MV) from which we started this proof.)

Let us concede this point immediately that (Strong Loc) is not an unqualified locality condition, and ask whether (Strong Loc) is at least a locality condition if we assume that all the various $mv_{ab}(a_n)$ and $mv(a_n)$ as well as $mv_{ab}(b_n)$ and $mv(b_n)$ exist. Clearly if all these mvs exist then failure of (Strong Loc) means that (for some **a**, **b**, n)

$$M_{ab} \to \sim a_n \quad \text{or} \quad M_{ab} \to \sim b_n.$$

(Note that we here regard '**a**', '**b**', 'n' as variables.) Suppose (with no loss of generality) that it is

$$M_{ab} \to \sim a_n.$$

Now abbreviate '**a** is measured on L_n at t' to 'a_n meas'. Hence, by definition of 'a_n',

$$a_n \text{ meas} \to a_n,$$

and hence we have

$$a_n \text{ meas} \to a_n \quad \text{and} \quad a_n \text{ meas.} \, M_{ab} \to \sim a_n.$$

But this together with Poss M_{ab} (which we have assumed) is easily shown to imply *both* that

(i) a_n meas. $\sim M_{ab} \to a_n$ and a_n meas. $M_{ab} \to \sim a_n$,

and that the counterfactuals in (i) are non-trivial, i.e. their respective antecedents are both possible.[10]

And it is easy to see that (i) involves a form of non-locality if we adopt the following traditional concept of causal relevance:

(CR) p is causally relevant to q if p is a necessary part of a sufficient condition for q, i.e., for some r, $r.p \to q$ and $r. \sim p \to \sim q$ non-trivially.[11]

---

[10] For convenience I prove the relevant inference using a Lewis semantics.

Proof Assume $p \to q$ non-trivially and $p.r \to \sim q$ non-trivially. Hence there is a sphere of worlds S_1 containing p-worlds so that within S_1 all p-worlds are q-worlds; and there is a sphere S_2 containing $p.r$ worlds so that all $p.r$ worlds within S_2 are $\sim q$-worlds. But clearly this is only possible if none of the p-worlds in S_1 is also an r-world; and hence all p-worlds within S_1 are $\sim r$-worlds. Hence all $p. \sim r$ worlds within S_1 are q-worlds. Moreover there must be some such $p. \sim r$-worlds. Hence $p. \sim r \to q$ non-trivially.

[11] Cf. McKie's definition of 'p is a cause of q' in McKie (1975). I follow Lewis (1975) p. 187 in taking this relation as one of causal *relevance*, rather than causation, because of its transitivity. Note also that without the non-triviality clause (CR) would have the undesirable consequence that any p is causally relevant to any q, by simply setting $r = p.q$.

That is from (CR) and (i) it follows that $\sim M_{ab}$ is causally relevant to a_n. And because of the large separation of the various electrons this implies an objectionable non-locality, *qua* the faster-than-light transmission of some causal effect (or, more correctly, it implies a causal relation between events at a time-like separation). Note however that this locality failure only has a metaphysical existence since the causal dependence of a_n on $\sim M_{ab}$ can never be used to affect a_n by switching from $\sim M_{ab}$ to M_{ab}. This is because it follows from (i) that $\sim M_{ab}$.[12] This result agrees with Redhead's result that 'Bell telephones' cannot be constructed (1986). Nevertheless we see that (Strong Loc) failure does imply non-locality if we assume that all the $mv_{ab}(a_n)$ exist. In particular we see that the previous section's result means that *if* we discount the possibility of indefinitely improbable coincidences and accept the laws of QT then we must admit either (1) objectionable non-localities or failure of freedom of measurement (FM) or (2) the non-existence of some $mv_{ab}(a_n)$ (or $mv_{ab}(b_n)$) or (3) the non-existence of some $mv(a_n)$ or $mv(b_n)$ (and hence reject (MV)).

But there is yet another response to the demonstration in section 1 of a form of contradiction between (FM), (Strong Loc), QT, and (MV). It is to deny that what we have demonstrated there even amounts to a 'pragmatic contradiction' between these propositions. I shall call this response, following Butterfield,[13] 'the ropey world objection'. In brief it accepts that the argument in section 1 shows that (at least) one of w_1, w_2, w_3, w_4—say w_1—must be 'ropey' in the sense that a statistical estimator, viz. $E^1(\mathbf{a}, \mathbf{b})$, for some quantity defined on w_1, viz. $\langle \mathbf{a} \times \mathbf{b} \rangle_F$, deviates significantly from what QT predicts that its value is, where *ex hypothesi* QT holds at w_1. More particularly if we let $(2 + 4c)$ be the value which QT predicts that the left-hand side of (i)' has then we see that $E^1(\mathbf{a}, \mathbf{b})$ must deviate from $\langle \mathbf{a} \times \mathbf{b} \rangle_F$ by at least c (or $E^2(\mathbf{a}, \mathbf{b}')$ must deviate from $\langle \mathbf{a} \times \mathbf{b}' \rangle_F$ by at least c, *or* . . .). And if N is large enough then such a deviation must be statistically significant because $P^1(|E^1(\mathbf{a}, \mathbf{b}) - \langle \mathbf{a} \times \mathbf{b} \rangle_F| > c)$ can be made as small as we like for N large enough.

But this conclusion (so the objection continues) is not a contradiction, pragmatic or otherwise, nor even surprising, because it is after all always possible for a statistical estimator to deviate significantly from

---

[12] *Proof* (by *reductio*): Suppose M_{ab} holds in the actual world, and hence so does 'a_n meas'. Hence 'a_n' also holds at the actual, and so 'a_n meas $M_{ab} \rightarrow \sim a_n$' is false at the actual—in contradiction with (i).

[13] J. Butterfield, private interchange.

the correct value of the quantity it estimates; indeed over a large enough number of applications there is even a high probability that such a deviation will occur sometimes. And the world w_1 can then be seen as a possible situation where precisely such a deviation occurs.

A modification of the argument of section 1 will circumvent this ropey worlds objection however. Let N go to infinity in the section 1 argument. A result by Hartle (1968) tells us that if N is infinite then $E^1(\mathbf{a}, \mathbf{b}) = \langle \mathbf{a} \times \mathbf{b} \rangle_F$; and the latter identity then transforms the alleged pragmatic contradiction demonstrated in section 1 into a formal contradiction.

Another less tendentious[14] response to the ropey worlds objection is to accept it, but then show that the existence of the ropey world itself generates further difficulties as follows. What we showed in section 1 is that for any particular N, however large, and for any particular choice of counterfactual measured values there is always one of the four possible apparatus settings $(\mathbf{a}, \mathbf{b})$, $(\mathbf{a}', \mathbf{b})$, $(\mathbf{a}, \mathbf{b}')$, or $(\mathbf{a}', \mathbf{b}')$—call it 'the pathological setting'—which would result in a deviation of the type described in the previous paragraph, although what that pathological setting is may well depend on N or the counterfactual measured values.

But, I shall now argue, the existence of this pathological setting, i.e. the existence of the ropey world, itself generates a contradiction with yet another metaphysical principle which it is plausible to assume. Consider the experimentally observed fact that deviations of the type described in the previous paragraph have not in fact occurred more often than we would expect if they had always been highly improbable in the actual world. This means that, by and large, measurement apparatus settings in the actual world have been correlated with the counterfactual measured values of the q-quantities for the systems they were measuring in such a way that pathological settings have not occurred.[15]

But how do we explain this last correlation? Reichenbach's Common-cause Principle implies that the correlated items are either directly causally related or share a prior cause. But it is precisely such

[14] Less tendentious because it neither assumes there is an infinite number of q-systems, nor needs the validity of QT in the limiting situations considered by Hartle.
[15] Or at least it is highly probable that this has been so, so probable in short that we are justified in taking this too as an 'experimental fact'. The probability here is the physical probability defined in QT at the actual world. It is the law of large numbers which enables one to use this probability to talk of the probability, in the actual world, of the deviations occurring.

causal restrictions on what is measured which (FM) rules out, i.e. (FM) entails that the values of the physical quantities do not restrict which of them can be measured. So we do derive a contradiction just from the existence of the ropey worlds, if we additionally assume the above Reichenbach principle. And so the ropey worlds objection simply shifts the problem: from having to reject one of (Strong Loc), (FM), (MV), and QT to rejecting one of these *or* the Reichenbach principle. It is of no help here, to make the point which I made in Chapter 3, that the Reichenbach principle is in doubt for correlations between indeterministic factors, because the factors here are not indeterministic. So we again face the three options (1) failure of (FM) or locality; (2) the non-existence of some $mv_{ab}(\mathbf{a}_n)$, and (3) failure of (MV).

I shall now argue that the best of these options is (3). First consider the options (2) and (3). One way of implementing them (either (2) or (3)) is by allowing the relevant measured values to depend on slight variations of the measurement context which do not change the relevant measurements' distance from the actual world, so that two equally nearest possible measurements have different results and hence a unique mv does not exist. This can be achieved within a hv deterministic programme simply by introducing new variables which constitute the 'missing context' (missing from the point of view of QT) and on which the mvs depend. Yet another even simpler way of implementing these options is to take the measurement process as indeterministic, as discussed in Chapter 7 and by Redhead (1983) and Hellman (1983b). Or one could appeal to the existence of an intrinsic probability distribution over the possible measurement outcomes, while carefully distinguishing this from indeterminism, as I have done in Chapter 7.

Of the options (2) or (3), (3) is then the more plausible because the measurements considered in (3) are less restrictively specified, i.e. we do not specify M_{ab} as part of the measurement context in (3), and this makes it more credible that there should be a variation over the results of the re-runs of the measurements considered in (3) than for the corresponding measurements considered in (2). In short it seems implausible that specifying the context of measurement more accurately (as we do in (2)) should have the consequence that repeat measurements become less uniquely specified in their results (than for (3)). (Although there is of course no *logical* inconsistency in supposing this might happen. It is the failure of the rule '$p \rightarrow q$ ∴ $p.r \rightarrow q$' which makes this a possibility.) Moreover the option (3), implemented in any of the ways just discussed, seems clearly preferable to option (1) with its violation of a priori principles. In short (3) emerges as the preferred option.

In sum then the argument of section 1 showed that soft realism, *qua* (MV), is pragmatically inconsistent with (Strong Loc), (FM), and QT, and *a fortiori* that hard realism is too. This argument elicits various responses which we discussed in this section: modifying QT; rejecting (MV) by adopting a contextualized hv programme; or appealing to indeterminism, or an intrinsically probabilistic model for measurement; or the ropey worlds objection to the argument itself (questioning the 'inconsistency'); or even of course questioning one or more of the classical metaphysical principles (FM) or (Strong Loc). Of these responses I suggest that it is the giving up of soft realism, *qua* (MV), in the context of taking measurement processes as intrinsically statistical, which is most satisfactory. As indicated in Chapter 7, this response does not exact too heavy a metaphysical price: the failure of (MV) can be seen as simply an aspect of the intrinsically statistical nature of QT, which we argued for independently there.

The extension of this conclusion to the rejection of hard realism, and hence of either (Det Q) or (Pass Q) (in some form), must then be faced. And here my previous commitment to (Bohr)' commits me to the surrender of (Pass Q) in one of its forms (see Chapter 7) and in particular to the surrender of the view that any determinate possessed value is matched by a counterfactually construed measured value. To surrender this is of course consistent with holding all actual measurements to be passive, and even holding that all possible measurements are 'locally passive' in the sense that their mvs are always equal to the value that the measured quantity would have were a measurement to take place. To achieve the latter form of passivity however, we now see that we must allow that measurement may change what is measured even before the measurement proper starts, i.e. we must allow a 'pre-measurement interaction'. In other words we can still have, for any q,

(Loc Pass)   Meas $Q \rightarrow (\text{val } Q = q \text{ iff } \text{mv} Q = q)$,

without assuming that there is always a particular q for which

Meas $Q \rightarrow \text{mv} Q = q$.

3. THE LIMITS OF DE-OCKHAMIZATION

Some of the previous results may not seem much of an advance on those of Chapter 7 where, without assuming either (FM) or locality— and simply in the light of the Kochen and Specker theorem—it was shown that some passivity principles had to be surrendered if (Det Q) was to be preserved within the framework of QT (and similarly if the modified Bohr interpretation of QT was adopted). But the advantage of the results we have just proven is, as we shall see, that they apply even if we make the revisions of standard QMT which we suggested in Chapter 7 in order to preserve *both* (Det Q) and passivity principles in the light of KS.

We pointed out in Chapter 7 that the de-Ockhamizer could (as far as the results of that chapter were concerned) save both (Det Q) and all of (Pass Q) by restricting the measurements of Q to measurements of a non-degenerate Q of which Q is a function (van Frassen's anti-Copenhagen strategy (1973) p. 107). But I shall now show that even if we make the latter restriction the results we have just derived show that passivity cannot be saved, at least not if we are to preserve (FM) and avoid failures of locality. The proof here turns on an ambiguity in the notion of 'degenerate' which it is important to expose. Suppose at time t we measure **a** for S_1, the L-electron of some well-separated pair of electrons in state F, with S_2 as the other member of the pair. And suppose we measure **a** by measuring some properly de-Ockhamized physical quantity for S_1—call it '$Q(\mathbf{a})$'—of which **a** is a function. This $Q(\mathbf{a})$ is supposed to be de-Ockhamized in the sense that the operator representing it in QT represents no other physical quantity; and hence it can be measured unequivocally by the techniques of standard QMT (which measures a physical quantity in terms of methods sensitive only to the representative operator for that quantity). The de-Ockhamization introduced in Chapter 7 took it that all and only those physical quantities for S_1 which are non-degenerate in the sense that they are represented by non-degenerate operators on $\mathscr{H}(S_1)$ are de-Ockhamized in the relevant sense. Thus $Q(\mathbf{a})$, if it is to be de-Ockhamized, must be non-degenerate in the sense that the operator on $\mathscr{H}(S_1)$ representing $Q(\mathbf{a})$ is non-degenerate (unlike the operator on $\mathscr{H}(S_1)$ representing **a**, which happens to be degenerate according to QT).

The crucial assumption then is that the $Q(\mathbf{a})$ measuring device (described within QMT) is also localized near S_1 at t, just as the **a**-

detector was in our earlier example, so that the measurement of **a** by measuring $Q(\mathbf{a})$ at t is not affected by any processes localized in the vicinity of S_2. And this is plausible because $Q(\mathbf{a})$, just like **a**, is conceived of as a physical quantity characterizing S_1. And the same goes for the apparatus used to measure $Q(\mathbf{b})$ at t, i.e. it is restricted to the vicinity of S_2 at t.

In this way we see that the restriction to measuring functions of de-Ockhamized physical quantities makes no difference to the argument of the previous section. We still prove that in order for both QT and (FM) to obtain either (MV) fails—in particular the mv of **a** by the method of measuring $Q(\mathbf{a})$ does not exist for some **a** and some L-electron in C at t—or if all such mvs exist then (Strong Loc) fails.

The de-Ockhamizer may object to the preceding argument as follows. $Q(\mathbf{a})$ is not non-degenerate after all, even though it is represented by a non-degenerate operator—let it be $\hat{Q}(\mathbf{a})$—on $\mathscr{H}(S_1)$. In support of this he will point out that $Q(\mathbf{a})$ is represented by a degenerate operator on $\mathscr{H}(S_1 + S_2)$, viz. by $\hat{Q}(a) \times \hat{I}_2$, where $\hat{I}_2$ is the identity operator on $\mathscr{H}(S_2)$. As such a measurement of $Q(\mathbf{a})$ by the techniques of QMT cannot be relied on to produce a unique mv, but may instead yield different values depending on which of the properly de-Ockhamized physical quantities represented by $\hat{Q}(\mathbf{a}) \times \hat{I}_2$ is being measured. In this way we can explain the failure of (Strong Loc) as follows. What appears to be a case where the mv of **a** in S at t is dependent on whether **b** or **b**' is measured at t on S_2 is really a case where two different de-Ockhamized physical quantities are measured by two different methods—one by measuring $Q(\mathbf{a})$ in the context of measuring **b** on S_2 and the other by measuring $Q(\mathbf{a})$ in the context of measuring **b**' on S_2.

But there is a serious problem for the de-Ockhamizer in postulating the extra aspect to de-Ockhamization that we have just introduced, viz. we can make any apparently non-degenerate physical quantity Q for S (represented in $\mathscr{H}(S)$ by the non-degenerate operator $\hat{Q}$) degenerate by considering it in the context of some broader system $S + S'$, and hence deny that it is fully de-Ockhamized. Moreover, since S' can be arbitrarily varied—as $S + S'$, $S + S' + S''$, $S + S' + S'' + S'''$, ... —there is no end to such requirements of further de-Ockhamization. Rather than reduce de-Ockhamization to absurdity in this way, the de-Ockhamizer would seem best served to agree that Q—a physical quantity for S (and *a fortiori* for $S + S'$, for any S')— is 'non-degenerate' in the relevant sense needed for it to be fully de-

Ockhamized if it is represented by a non-degenerate operator on $\mathscr{H}(S)$. That this same Q is also represented by degenerate operators on any $\mathscr{H}(S+S')$ is then set aside as irrelevant.

Accepting this, we see that we have indeed extended the results of Chapter 7. By introducing (FM) and a locality requirement (Strong Loc) we have shown that the realist's (Pass Q) and (Det Q) are not reconcilable with QT even if we revise QMT as the de-Ockhamizer suggests by restricting proper measurements to measurements of functions of non-degenerate physical quantities.

9

Heywood and Redhead

Eigenvector rule; Heywood and Redhead realism; Kochen and Specker; completeness; de-Ockhamization; Redhead and Heywood argument; OLOC; ELOC.

INTRODUCTION

Two varieties of realism in QT (Quantum Theory) are distinguished. A strong form which endorses:

(VR) If there is zero probability of the measured value of Q being q for system S at time t then Q does not have the value q for S at t;

and a weaker form in which what we call the 'eigenvector rule' is endorsed. ((VR) is seen as a realist principle because it establishes a mirroring relation between the reality of possessed values and what is measured.) It is argued that if we adopt a minimum completeness condition for QT, one which is ·already incorporated within the standard de-Ockhamization strategy, then the distinction between these two varieties of realism collapses. Moreover adapting an argument by Heywood and Redhead (1983) it is shown that the strong form of realism just defined is inconsistent with Heywood and Redhead's principle OLOC. But it is also argued that within a realist metaphysics OLOC emerges as a consequence of QT and certain 'analytic' principles which give meaning to physical quantities in sub-systems. These results enable a transformation of the Heywood–Redhead argument into an argument against combining either of the above two forms of realism with a standardly de-Ockhamized QT. (By contrast Heywood and Redhead present their argument as merely showing that realist interpretations of standardly de-Ockhamized QT must involve locality failures of a rather quixotic variety.) These results represent a strengthening of Bell-type arguments which only purport to show that a *local* realist interpretation of QT is not possible. They also represent a strengthening of the Kochen and Specker arguments of Chapter 7, since they do not assume standard Quantum Measurement Theory (QMT).

1. DE-OCKHAMIZING AND (VR)

Heywood and Redhead conduct their proof against a background of certain core assumptions from QT which we now briefly revise. With every quantum system S is associated a Hilbert space $\mathscr{H}(S)$ the elements of which represent the possible states of S at particular times. With every discrete-valued physical quantity Q for S is associated a complete orthonormal set of eigenvectors in $\mathscr{H}(S)$, each eigenvector being associated with one of the possible values for Q and such that

(i) If S at t is in a pure state represented by the vector f in $\mathscr{H}(S)$ then f is one of the eigenvectors in $\mathscr{H}(S)$ associated with the possible value q_i iff Prob $(Q, q_i, S, t) = 1$,

where Prob (Q, q_i, S, t) is the probability of the measured value of Q being q_i on measuring Q in S at t. The possible values q_i are referred to as 'eigenvalues of Q'.[1] This principle (i) follows trivially from the Born Statistical Interpretation for QT. Note that the measurement referred to here is an ideal measurement as defined within standard QMT. But in this context the precise identity of the Prob (Q, q_i, S, t) does not concern us; it is only their formal role, in (i) for example, which is of relevance.

We also assume the eigenvector rule:

(ER)   If S at t is in the pure state f_i where f_i is an eigenvector of Q for eigenvalue q_i in $\mathscr{H}(S)$ then $Q(S, t)$ exists and is equal to q_i,

where $Q(S, t)$ is the value possessed by Q for S at t. Note that under the Copenhagen interpretation of QT a sort of converse of (ER) holds too, i.e. if S at t is in a pure state then Q has value q_i in S at t *only if* $f(S, t)$ (the

---

[1] Implicit here is what is usually called the 'spectrum rule', viz. that the eigenvalues of Q (defined as the eigenvalues of $\hat{Q}$) exhaust all the possible (or at least actual) values possessed by Q.

state-vector for S at t) is an eigenvector of Q in $\mathcal{H}(S)$.[2] But under a realist interpretation of QT this converse does not hold. Rather it is supposed that

(Det Q) $Q(S, t)$ exists for any S, t (t is here implicitly restricted to values during the lifetime of S).

Indeed it is this last assumption which (along with passivity of measurement to be introduced later) characterizes the 'realist' metaphysical framework for QT which is being set up here. We can also extend (ER) to:

(ER)*   If S at t is in f_i then $\text{Prob}(Q(S, t), q_i) = 1$,

where the latter probability is the probability of $Q(S, t)$ being q_i (although this can also be seen as a special case of (Bohr)', or even just as a consequence of (i) and the (Pass)' of Chapter 6).

In the special case for which the set of eigenvalues of Q corresponds one-to-one to the eigenvectors for Q in $\mathcal{H}(S)$, Q is said to be maximal for S. (Derivatively we can then talk of the operator $\hat{Q}$ on $\mathcal{H}(S)$ which represents Q as being 'maximal' or 'non-degenerate' on $\mathcal{H}(S)$. Note that I have here reversed the usual order of definition in taking maximality as in the first instance a relational property of physical quantities and systems, and only derivatively as a property of operators. This reversal of the usual order makes no difference to the formal statement of QT.) In that special case (i) can be strengthened to:

> S at t is in a pure state which is represented by the unique eigenvector of Q for eigenvalue q_i iff $\text{Prob}(Q, q_i, S, t) = 1$.

But more generally if Q is non-maximal for S then we cannot assume this; indeed all we can assume, and this is an extension of (i), is:

(i)' $\text{Prob}(Q, q_i, S, t) = 1$ iff S at t is in a mixture of eigenvectors of Q for eigenvalue q_i

(that mixture being pure in the special case that Q is maximal). We also assume, as a corresponding extension of (ER):

(ER)'   If S at t is in a mixture of eigenvectors of Q for eigenvalue q_i then $Q(S, t)$ exists $= q_i$.

---

[2] In fact the Copenhagen interpretation only requires that if $f(S, t)$ is not one of the eigenvectors of Q then the value of Q in S at t is indeterminate (rather than has no value at all). For this distinction see Chapter 1.

From (i) and (ER), in the case that S at t is in a pure state, or more generally from (i)$'$ and (ER)$'$, it then trivially follows that

(VR)$'$ If Prob $(Q, q_i, S, t) = 1$ then $Q(S, t) = q_i$.

Finally we assume:

(ii) $\sum_{q_i}$ Prob $(Q, q_i, S, t) = 1$.

(Note however that we cannot write:

$$\sum_i \text{Prob } (Q, q_i, S, t) = 1,$$

because for any Q which is non-maximal for S a given Prob (Q, q_i, S, t) may be counted twice in the summation.) This (ii) also follows trivially from the Born Statistical Interpretation.

We now introduce some new probabilities. If I is an interval of possible values for Q, then we define:

$$\text{Prob } (Q, I, S, t) = \sum_{q_i : q_i \in I} \text{Prob } (Q, q_i, S, t)$$

$$\text{Prob } (Q, \bar{I}, S, t) = \sum_{q_i : q_i \notin I} \text{Prob } (Q, q_i, S, t)$$

And trivially (from (ii)) we then see that

(ii)$'$ Prob $(Q, I, S, t) + $ Prob $(Q, \bar{I}, S, t) = 1$.

The question now is the relation of (VR)$'$ with:

(VR)   If Prob $(Q, q_i, S, t) = 0$ then it is not the case that $Q(S, t) = q_i$.

It is easy to prove that (Det Q) and (VR) and (ii) together entail (VR)$'$:

Proof   Let Prob $(Q, q_i, S, t) = 1$. Then, by (ii), Prob $(Q, q_j, S, t) = 0$ for any $j \neq i$. Hence, by (VR), it is not the case that $Q(S, t) = q_j$ for any $j \neq i$. But by (Det Q), $Q(S, t) = q_j$ for some j (since the set of q_j for varying j is defined as the set of possible values for Q).[3] Hence $Q(S, t) = q_i$. QED.

But the converse implication from (Det), (VR)$'$, and (ii) to (VR) does not hold, and in that respect (VR) represents a strengthening of (VR)$'$. (Although later I shall show that in the context of assuming a minimal

---

[3] Had we taken the various q as eigenvalues of the operator $\hat{Q}$ representing Q in the usual way then we would need to appeal to the 'spectrum rule' here, that the only values Q takes are its eigenvalues.

completeness condition for QT and the Born Statistical Interpretation the converse implication does hold.) The reason for the asymmetry just pointed out (i.e. that (VR), (Det Q), and (ii) entail (VR)′ whereas (VR)′, (Det Q), and (ii) do not entail (VR)), is the following further asymmetry. If a set of probabilities such as the Prob (Q, q_i, S, t) sum to 1 then from one of them taking the value 1 we can infer that the others take the value 0 (it is this inference which we used to deduce (VR)′ from (VR), (Det Q), and (ii)); but we cannot infer conversely from one of the $P(Q, q_i, S, t)$ taking value 0 to one of the others taking the value 1. Note that this last asymmetry does not depend on any of the non-classical features of QT, but is purely classical in origin.

The asymmetry between (VR) and (VR)′ just pointed out is still apparent when we consider generalized versions of them, viz.:

(GVR) If Prob $(Q, I, S, t) = 0$ then it is not the case that $Q(S, t) \in I$.

(GVR)′ If Prob $(Q, I, S, t) = 1$ then $Q(S, t) \in I$.

(GVR) is easily seen to be equivalent to (GVR)′ given (Det Q) and (ii). And similarly it is easy to show that (GVR)′ is equivalent to (VR). But whereas (GVR) is a trivial variant on (VR) this is not so for (GVR)′ and (VR)′. The reason for this asymmetry is exactly parallel to the reason for the asymmetry discussed in the previous paragraph, viz. that from Prob $(Q, I, S, t) = 1$ we cannot conclude that Prob $(Q, q_i, S, t) = 1$ for some $q_i \in I$, whereas we can conclude from Prob $(Q, I, S, t) = 0$ that Prob $(Q, q_i, S, t) = 0$ for all $q_i \in I$. And as such the inference of (GVR)′ from (VR)′, unlike that of (GVR) from (VR), requires some extra principle such as (VR). This asymmetry too does not rely on any of the non-classical features of QT, or on adopting any particular interpretation of QT. However the asymmetry is particularly clear in the context of the Copenhagen interpretation, because whereas the latter interpretation rejects (GVR)′ it accepts (VR)′. For example it allows that Prob $(Q, (-\infty, \infty), S, t) = 1$ whereas $Q(S, t)$ may not exist at all, i.e. it agrees that the measured value of Q lies somewhere in the interval $(-\infty, \infty)$ even though there may be no determinate value for Q to have.

This distinction between (VR) and (VR)′ is a distinction which I shall later argue collapses if we adopt an uncontroversial passivity assumption and the standard de-Ockhamizing strategy. Indeed this demonstration is one of the two key moves which I use to strengthen the Heywood–Redhead result. For now however I shall persevere with the distinction between (VR) and (VR)′ for the sake of generality.

Heywood and Redhead in (1983) work within the realist framework for QT which I have just developed above, although they generalize QT by weakening (VR) to:

(VRR) If $\text{Prob}(Q, q_i, S, t) = 0$ then it is not the case that $Q(S, t)(Q) = q_i$,

where $Q(S, t)(Q)$ is the value possessed by Q in S at t when S at t has been prepared for a Q-measurement. In Appendix 12 I discuss what difference this weakening of (VR) makes, but in the body of the text I work with (VR) rather than (VRR).

I shall now summarize the consequences of introducing the Kochen and Specker theorem within the above realist framework for QT. What I say here will overlap what was said in Chapter 7, but will come at matters from a somewhat different perspective. It is convenient first of all to introduce the notion of the operator representing a physical quantity. If Q has eigenvectors $\{f_i\}$ on $\mathscr{H}(S)$ for corresponding eigenvalues $\{q_i\}$ then the operator $\hat{Q}$ on $\mathscr{H}(S)$ which is said to represent Q (on $\mathscr{H}(S)$) is linear self-adjoint and satisfies the eigenvalue equation:

$$\hat{Q}f_i = q_i f_i \text{ for all } i.$$

(Since $\{f_i\}$ is *ex hypothesi* complete and orthonormal this last set of equations uniquely specifies $\hat{Q}$, as we discuss in Appendix 1.)

Particularly important examples of physical quantities represented by operators are two-valued physical quantities which have possible values 1 and 0 with just the one eigenvector for the eigenvalue 1. These are represented by one-dimensional projection operators ('projectors' for short) q.v. Let f_1 be the unique eigenvector in $\mathscr{H}(S)$ of the two-valued physical quantity P in $\mathscr{H}(S)$ for eigenvalue 1, and $\{f_2, f_3, \ldots\}$ be the eigenvectors in $\mathscr{H}(S)$ of P for the other eigenvalue 0. Then, letting $\hat{P}$ be the operator on $\mathscr{H}(S)$ representing P, we have that

$$\hat{P}f_1 = f_1 \text{ and } \hat{P}f_i = 0 \text{ for any } i \neq 1.$$

It is then easy to show that $\hat{P}$ so specified is the one-dimensional projector onto f, i.e. $\hat{P}$ satisfies:

$$\hat{P}f = (f_1, f)f_1$$

(because the $\{f_i\}$ is complete and orthonormal in $\mathscr{H}(S)$).

The KS theorem, as presented in Chapter 7, was used to show a contradiction within a realist framework when it is supplemented by the following assumptions:

(C1) For any system S, for any Q which is maximal for S, and any eigenvalue q_i of Q, there is a q-quantity Q_i for S which satisfies:

$Q_i(S, t) = 1$ if $Q(S, t) = q_i$, and Prob $(Q_i(S, t), 1)$
$\quad = 1$ if Prob $(Q(S, t), q_i) = 1$.

$Q_i(S, t) = 0$ if $Q(S, t) = q_j$, for any $j \neq i$, and Prob $(Q_i(S, t), 0) = 1$ if Prob $(Q(S, t), q_j) = 1$ for any $j \neq i$.

(C2) Any physical quantity which is represented by the same operator, i.e. the same eigenvalues and eigenvectors, as Q_i has the same values as Q_i at all times. (This is just a rewrite of (U) of Chapter 7.)

(C1) can effectively be seen as a rather minimal completeness condition for QT. And the uniqueness condition (C2) can also be seen as a completeness condition, i.e. it is a special case of the claim that if two physical quantities are represented in the same way in QT then they are the same at least in respect of having the same values. In other words where QT makes no distinction in representation there is no difference in reality. I shall now prove that KS does indeed generate a contradiction between 'realism', construed in terms of the preceding principles of this chapter, and the completeness conditions (C1) and (C2). This constitutes a more precise reworking of the proof already given in Chapter 7.

From (C1) we can prove that the operator $\hat{Q}_i$ representing Q_i is $\hat{P}(f_i)$, the projection operator onto f_i (the eigenvector of Q for eigenvalue q_i) by introducing the following uncontroversial minimal passivity of measurement condition (introduced in Chapter 6):

(Pass Q)′ Prob $(Q, q_i, S, t) = 1$ iff Prob $(Q(S, t), q_i) = 1$.

This passivity condition says that at least in the extreme case where the probability of measuring Q to have value q_i is 1 this probability is equal to the probability of Q possessing the value q_i, and vice versa. This condition is one which even exponents of the Copenhagen interpretation would accept (although they would not accept it for probabilities less than 1).[4]

---

[4] I.e. for Copenhageners Prob $(Q, q_i, S, t) = 1$ guarantees that S at t is in a pure state (or a mixture of such) which is an eigenvector of Q for value q_i and it is precisely under those conditions, and only those conditions, that it makes sense to assign Q the value q_i in S at t. Note however that some positivistic Copenhageners may additionally require that Q is actually measured in S at t, not just that S at t is prepared to be (or has simply evolved to be) in an eigenvector of Q for eigenvalue q_i.

Proof Let $f(S, t) = f_i$. Then Prob $(Q, q_i, S, t) = 1$ (by (i)). Hence Prob $(Q(S, t), q_i)$ $= 1$, by (Pass Q)'. (Or we can go directly from $f(S, t) = f_i$ to Prob $(Q(S, t), q_i) = 1$ via (ER)*.) Hence, by (C1), Prob $(Q_i(S, t), 1) = 1$; and so, by (i), $f(S, t)$ is an eigenvector of Q_i for eigenvalue 1. By letting $f(S, t)$ $= f_j$ we can then similarly prove that, for any $j \neq i, f_j$ is an eigenvector of Q for eigenvalue 0. But the only linear self-adjoint operator with such eigenvectors and eigenvalues is $\hat{P}(f_i)$. QED.

We can now see how (C1) and (C2) generate a contradiction within the realist framework given above. Consider a Q which is maximal for S with eigenvalues $\{q_i\}$ and $\{f_i\}$ as the eigenvectors of Q in $\mathcal{H}(S)$. By (C1) we can suppose Q_i exists for each q_i with $\hat{Q}_i = \hat{P}(f_i)$ and $Q_i(S, t) = 1$ if $Q(S, t) = q_i$ and $Q_i(S, t) = 0$ if $Q(S, t) = q_j$ for any $j \neq i$. If we assume (Det Q) then, for some i, $Q(S, t) = q_i$, and hence $Q_i(S, t) = 1$ and $Q_j(S, t)$ $= 0$ for any $j \neq i$. Thus

$$\sum_i Q_i(S, t) = 1 \text{ and } Q_i(S, t) = 1 \text{ or } 0 \text{ for all } i.$$

Moreover if two such maximal Q, Q' have an eigenvector in common, say $f_i = f'_j$ so that $\hat{P}(f_i) = \hat{P}(f'_j)$, then $Q_i(S, t) = Q'_j(S, t)$, by (C2).

Now define a map m on the domain of vectors which are eigenvectors of the various maximal Q such that:

$$m(f_i) = Q_i(S, t).$$

Since for $f_i = f'_j$ it is the case that $Q_i(S, t) = Q'_j(S, t)$ it follows that m is single-valued. Moreover $\Sigma_i m(f_i) = 1$ and $m(f_i) = 1$ or 0 for all i for any $\{f_i\}$ which are eigenvectors of such a maximal Q. But it is well known that no such map m can exist if the domain of m is large enough, i.e. if there are enough of the relevant maximal Q and if $\mathcal{H}(S)$ is at least three-dimensional. Moreover it is easy to show that there is such an S, i.e. let S be a spin-1 system (e.g. an electron pair), and the various Q be various spin Hamiltonians of the form $aS_x^2 + bS_y^2 + cS_z^2$ for the 117 different combinations of directions x, y, z described by Kochen and Specker in their (1967) (S_x^2 is the square of the spin component of S along the x direction, etc.). QED.

It is perhaps appropriate also to revise here the more usual way of looking at (C1) and (C2), viz. to focus on the claim which (C1) and (C2) make that the functional relation between Q_i and Q is identical with the functional relation between $\hat{Q}_i$, viz. $\hat{P}(f_i)$, and $\hat{Q}$. We can see how they make this claim by introducing a function D_q defined by

$$D_q(x) = 1 \text{ if } x = q$$
$$= 0 \text{ if } x \neq q.$$

Then clearly

$$Q_i(S, t) = D_{q_i}Q(S, t),$$

which we may abbreviate simply to:

$$Q_i = D_{q_i}(Q).$$

(Note that I am here taking the statement that $Q = Q'$ as shorthand for the statement that $Q(S, t) = Q'(S, t)$ for all t. In particular $Q = Q'$ should not be taken to imply that Q and Q' are identical in any stronger intensional sense. In the same way we may write the gas law '$P = nRT/V$' without implying that the pressure of a gas is the same concept as the product of the number of moles of the gas with its temperature divided by its volume, in some appropriate set of units.)

But we also have

$$\hat{P}(f_i) = D_{q_i}(\hat{Q}).$$

This follows from the familiar definition of functions of operators, i.e. if $\hat{Q} = \Sigma_j q_j \hat{P}(f_j)$ then

$$D_{q_i}(\hat{Q}) = \sum_j D_{q_i}(q_j)\hat{P}(f_j).$$

As such from the uniqueness clause (C2) we see that for any Q' which is represented by $D_{q_i}(\hat{Q})$ it is also the case that $Q' = D_{q_i}(Q)$. Hence (C1) and (C2) can in part be seen as instances of the more general principle:

(Func)   if $\hat{Q}' = F(\hat{Q})$ then $Q' = F(Q)$,

which claims a functional isomorphism between the algebra of physical quantities, construed as a system of relations between their values, and the algebra of their representative operators; and this in turn can be seen as some sort of simplicity condition, viz. that the form of representation adopted for physical quantities is no more complex than is needed to reflect their functional relations.

Whichever way we look at (C1) and (C2) however, be it as completeness conditions or simplicity conditions, the point remains that they entail a contradiction with the above realist framework for QT, provided only that the set of physical quantities Q is large enough for some S for which the dimension of $\mathcal{H}(S) \geq 3$.

Note that this derivation of a contradiction can be carried through by starting from assumptions which are weaker than (C1) and (C2). In particular we can start from the assumption (Func) specialized to the case where the $\hat{Q}'$ are one-dimensional projectors on $\mathcal{H}(S)$ and the Q

are maximal for S—call this special one-dimensional version of (Func) '(Func 1)'.[5] Now again consider a system S, one for which not only various maximal Q exist, but also various Q_i exist where the Q_i are now defined simply by their representative operators being $\hat{P}(f_i)$ (rather than by the relation $Q_i = D_{q_i}(Q)$ as above). Then by (Func 1) it follows that $Q_i = D_{q_i}(Q)$, and the proof then continues as above provided only that enough of the Q, Q_i exist for S. And the spin-1 case considered by KS again provides just such a case, i.e. simply let the various Q be the various 117 spin Hamiltonians as above, and let $I - S_x^2$ (which is the projector onto the 0 value eigenvector for S_x) for the various x be the various Q_i. (This choice for Q, Q_i is possible because the 0 value eigenvector of S_x is also the eigenvector of $aS_x^2 + bS_y^2 + cS_z^2$ for eigenvalue $b + c$.)

This second proof may seem to be preferable to the first proof, given that it starts from a weaker assumption (viz. (Func 1) rather than (C1) and (C2)). But although (C1) and (C2) together are stronger than (Func 1) it is only in respect of the (in any case) highly plausible completeness condition (C1). Moreover the first proof has the advantage that the premises conveniently divide into (C1) and (C2), which are not separated out within (Func 1). So when we end up with a contradiction at the end of the first proof it is clear what we have to reject from the initial premises, viz. the uniqueness clause (C2). But this is not so clear in the case of the second proof; and for these reasons we persevere with (C1) and (C2) rather than (Func 1) as a starting-point for the KS argument.

We now summarize our conclusions from the KS argument. We see that both (Func), and more particularly (Func 1), must be surrendered by the realist; and more particularly if the realist wishes to retain the completeness of QT in the form of (C1) then (C2) must be surrendered. More generally he must surrender:

(U)   For all Q, Q' on S if $\hat{Q} = \hat{Q}'$ on $\mathcal{H}(S)$ then $Q = Q'$.

The latter strategy, of keeping (C1) and giving up (U) and more particularly (C2), is called 'de-Ockhamization'.[6] It means that for

---

[5] That (Func 1) is implied by (C1) and (C2) is obvious. It is because the only one-dimensional projectors $\hat{Q}'$ which are functions of $\hat{Q}$ are those for which, for some q_i, $\hat{Q}' = D_{q_i}(\hat{Q})$. Note that (Func 1) differs from the (Func)$_1$ of Chapter 7, which I shall have no recourse to in this chapter.

[6] As proposed by van Frassen (1973).

non-maximal physical quantities, like the Q_i,[7] physical quantities which are represented by the same operators may not always have the same values. More particularly there is always (at any time t) at least one projection operator on $\mathcal{H}(S)$ which represents two distinct physical quantities, distinct in that they have different values (at t).

De-Ockhamization need not however spread to maximal physical quantities; and I shall follow common usage in taking de-Ockhamization to preserve:

(U)* For maximal Q, Q' on S, if $\hat{Q} = \hat{Q}'$ on $\mathcal{H}(S)$ then $Q = Q'$

as well as preserving that part of (Func) which says:

(Func)* If $\hat{Q} = f(\hat{A})$ for some maximal A then there is some physical quantity—call it '$Q_{|A}$'—which is represented by $\hat{Q}$ and for which $Q_{|A} = f(A)$, i.e. $Q_{|A}(S, t) = f(A(S, t))$.

This (Func)* (which Heywood and Redhead endorse) clearly includes (C1).

De-Ockhamization construed in this way has radical consequences for measurement in QT as we indicated in Chapter 7, i.e. in standard QMT the measurement of a physical quantity Q for the system S is a function of its representative operator for Q, in that if $\hat{Q} = \hat{Q}'$ then any particular Q-measurement is also counted as a Q'-measurement and is taken to deliver the same measured value for Q as for Q'. But now consider such a pair Q, Q' for which, in accord with the failure of (U), $Q(S, t) \neq Q'(S, t)$. (Given (U)*, these Q, Q' must be non-maximal.) A measurement of both of them by the same measurement on S at t must fail to report correctly either the value of Q or Q', i.e. the value registered by the measurement cannot agree with both $Q(S, t)$ and $Q'(S, t)$ since these are *ex hypothesi* unequal. Hence measurement, even of the idealized variety defined in QMT, cannot be passive, i.e. faithfully report possessed values. Or, to make the same point differently, the standard notion of QMT that measurement of Q is a function of $\hat{Q}$ must be somehow revised if the principle of passive measurement is to be retained. Note however that we have only shown that this must be so for non-maximal physical quantities, since it is only these which we have shown must be de-Ockhamized.

One conservative policy which we could adopt in the light of these radical consequences for QMT is to assume that at least for maximal

---

[7] The Q_i are always non-maximal in more than two-dimensional spaces.

physical quantities QMT does specify ideal measurement procedures which are passive, i.e. we assume:

(Pass Q)* $\mathrm{mv}(A, S, t) = A(S, t)$ for maximal A, where $\mathrm{mv}(A, S, t)$ is the value which A would be measured to have were it measured in S at t.

From (Func)* and (Pass Q)* it then follows that even the non-maximal $Q_{|A}$ will be measured passively if we measure it by measuring the maximal A and then simply taking F of the resultant measured value. In other words if we denote the measured value of $Q_{|A}$ which is obtained in this way by '$\mathrm{mv}_A Q_{|A}$' then

$$\mathrm{mv}_A Q_{|A} = \mathrm{val}\, Q_{|A}.$$

Thus if we want to preserve passivity for non-maximal physical quantities then we can do so by restricting measurements to measurements of functions of maximal physical quantities, which work by measuring a maximal physical quantity and then taking a function of the measured value. I shall call this restriction of QMT 'conservative de-Ockhamization' ('conservative' because it preserves passivity). This is essentially the strategy adopted by van Frassen under the label 'anti-Copenhagen' (see (1973) p. 107).

I shall now show that (C1)—and *a fortiori* de-Ockhamization (conservative or not)—entails that the distinction between (VR) and (VR)' collapses if we assume the uncontroversial (Pass Q)'. This only involves proving that, given (C2), (VR)' implies (VR) (since we have already proved the converse above).

Proof Suppose $\mathrm{Prob}(Q, q_i, S, t) = 0$ for maximal Q. By (C1) a Q_i exists for which

(a) $Q_i(S, t) = 1$ if $Q(S, t) = q_i$

$Q_i(S, t) = 0$ if $Q(S, t) = q_j$ for $j \neq i$;

and as we showed above, using (Pass Q)', it follows that $\hat{Q}_i = \hat{P}(f_i)$ where f_i is the eigenvector of Q for eigenvalue q_i. But the Born statistical interpretation implies that

$$\mathrm{Prob}(Q, q_i, S, t) = \mathrm{Prob}(Q_i, 1, S, t).[8]$$

Hence $\mathrm{Prob}(Q_i, 1, S, t) = 0$; and so, by (ii) above, $\mathrm{Prob}(Q_i, 1, S, t) = 1$. Hence, if we assume (VR)', $Q_i(S, t) = 0$; and so (by (a)) it is not the case that $Q(S, t) = q_i$. QED.

[8] Because it implies both that $\mathrm{Prob}(Q, q_i, S, t) = \mathrm{Tr}\, \hat{W}(S, t)\hat{P}(f_i)$ and that $\mathrm{Prob}(Q_i, 1, S, t) = \mathrm{Tr}\, \hat{W}(S, t)\hat{P}(f_i)$ too.

So we conclude that if we assume standard QT together with the uncontroversial (Pass Q)′ then de-Ockhamization, or indeed just (C1), entails that the distinction between a weak realism (with just (VR)′) and a strong realism (with (VR)) collapses. (VR) can it seems be taken as a consequence of a realist interpretation of standard QT just in virtue of the uncontroversial (VR)′ (or equivalently the eigenvector rule (ER)). Thus we have bypassed the need to introduce Heywood and Redhead's (VRR) as a separate postulate. We have done so by starting from the uncontroversial (VR)′ (or (ER)), (Pass Q)′, and (C1), and hence deriving (VR).

2. THE HEYWOOD–REDHEAD ARGUMENT

Bell, Stapp, and Eberhard have all put forward arguments in various forms which show that a realist interpretation of QT, even a weak one which does not assume (VRR), is not possible without violating locality assumptions.[9] However their proofs can be criticized at various points; in particular for taking long-run relative frequencies as estimates of probabilities. Heywood and Redhead have produced a more rigorous argument to the same conclusion which bypasses this particular criticism; however this rigour comes at a price. The 'locality violation' which they demonstrate is that (at least) one of two principles which they call 'OLOC' and 'ELOC' respectively must fail. However they also seem to claim that it is only the violation of *both* of these principles which constitutes a 'proper' violation of locality (although the failure of OLOC is taken by them to be at least anomalous in that it entails a surprising non-classical form of 'holism' for physical quantities in joint systems—see Heywood and Redhead (1983) p. 498). Moreover the form of realism presupposed by Heywood and Redhead includes (VRR), and in that respect is stronger than the form of realism presupposed by Bell *et al.* So prima facie the Redhead and Heywood result seems to be of little significance: there is it seems no proof of locality failure as such, and the form of realism they consider contains a principle (viz. (VRR)) the loss of which would seem of no great moment to the realist.

In the rest of this chapter I shall argue to the contrary that the Heywood–Redhead (HR) argument can be used to provide a

---

[9] See the summary in Clauser and Shimony (1978).

powerful indictment of realism. In this section I shall simplify the HR argument to show that OLOC by itself (without ELOC) is in contradiction with a realist interpretation of QT which includes (VR). Then in the third section I shall argue that OLOC is a consequence of certain meaning postulates for the definition of physical quantities in the context of QT. Putting these results together with the results of the first section we conclude that it is realism itself (rather than just the disjunction of locality or realism) which is refuted by the HR argument, where the relevant 'realism' need only be strong enough to include (Det Q), the uncontroversial (Pass Q)' and the eigenvector rule (or (VR), or even just (VR)' which can be expanded into (VR) as indicated in the previous section) provided that we assume (C1) (which is in any case incorporated in the standard de-Ockhamization strategy). In Appendix 12 I shall then comment on what happens when the HR argument is applied to a weakened form of realism, with (VRR) instead of (VR), as HR do. In all these arguments de-Ockhamization of some form or other (although not necessarily conservative de-Ockhamization) will be assumed. I shall also, for ease of reading, drop the 'S, t' indices and render '$Q(S, t)$' more simply as 'val Q'.

Firstly I shall show that HR's principle OLOC, together with (VR), is in contradiction with the realist principle (Det Q) given standard QT. HR's principle OLOC is:

OLOC   If S_1 and S_2 are spatially separated and A is maximal for S_1 then for any X, Y maximal on $S_1 + S_2$, val $A \times I_{|X} =$ val $A \times I_{|Y}$,

where $A \times I_{|X}$ is that physical quantity for $S_1 + S_2$ which is a function of X and is represented by the operator $\hat{A} \times \hat{I}$ on $\mathscr{H}(S_1 + S_2)$, and $\hat{A}$ is the operator representing A on $\mathscr{H}(S_1)$, and $\hat{I}$ is the identity operator on $\mathscr{H}(S_2)$; and similar conventions apply to $A \times I_{|Y}$. It is of course implicit here that $\hat{A} \times \hat{I}$ is a function of both $\hat{X}$ and $\hat{Y}$, the representative operators of X and Y respectively on $\mathscr{H}(S_1 + S_2)$.

(*Note.* A is sometimes referred to as being *locally* maximal in that although it is maximal for S_1 it is not maximal for $S_1 + S_2$, i.e. although $\hat{A}$, an operator on $\mathscr{H}(S_1)$, is non-degenerate, $\hat{A} \times \hat{I}$ is degenerate on $\mathscr{H}(S_1 + S_2)$. Indeed any physical quantity will only ever be locally maximal since it is always possible to consider it as a physical quantity in a more comprehensive system.)

The first part of the proof (which follows exactly the steps in HR

(1983), section 2) consists in showing that from (VR) and (Func)* we can derive a generalization of (VR) to several physical quantities:

(CVR) For any Q, Q', R, physical quantities for S for which R is maximal in S and for which there are functions f, f' for which $\hat{Q} = f(\hat{R})$ and $\hat{Q}' = f'(\hat{R})$, if $\text{Prob}((Q, Q'), (q, q')) = 0$ then either val $Q_{|R} \neq q$ or val $Q'_{|R} \neq q'$,

where $\text{Prob}((Q, Q'), (q, q'))$ is the joint conditional probability of Q being measured to have value q and Q' being measured to have value q' conditional on both Q and Q' being measured in S at t.

Proof $\text{Prob}((Q, Q'), (q, q')) = \text{Tr} \, (\hat{W} \hat{P}(Q, q) \, \hat{P}(Q', q'))$,

where $\hat{W}$ is the density operator for S at t and $\hat{P}(Q, q)$ is the projection operator onto the eigenspace of Q in $\mathscr{H}(S)$ for eigenvalue q. Since $\hat{Q} = f(\hat{R})$ we have that

$$\hat{P}(Q, q) = \sum_{i: f(r_i) = q} \hat{P}_i, \text{ and } \hat{P}(Q', q') = \sum_{i: f'(r_i) = q'} \hat{P}_i,$$

where $\hat{P}_i$ is the projection onto f_i the eigenvector of R for eigenvalue r_i. Hence since $\hat{P}_i \hat{P}_{i'} = 0$ for $i \neq i'$, and $\hat{P}_i \hat{P}_i = \hat{P}_i$ the above probability is

$$\text{Tr} \, \hat{W} \sum_{i: f(r_i) = q \text{ and } f'(r_i) = q'} \hat{P}_i,$$

which is $\text{Tr} \, \hat{W} \sum_{i: r_i \in f^{-1}(q) \cap f'^{-1}(q')} \hat{P}_i$

$$= \text{Prob}(R, f^{-1}(q) \cap f'^{-1}(q)),$$

Now assume $\text{Prob}((Q, Q'), (q, q')) = 0$. Hence $\text{Prob}(R, f^{-1}(q) \cap f'^{-1}(q')) = 0$, which, by (VR), means that val $R \notin f^{-1}(q) \cap f'^{-1}(q')$. Hence either val $R \notin f^{-1}(q)$ or val $R \notin f'^{-1}(q')$. That is either $f(\text{val } R) \neq q$ or $f'(\text{val } R) \neq q'$, which, by (Func)*, means that val $Q_{|R} \neq q$ or val$_{|R} Q' \neq q'$. QED.

We now apply (CVR) to a special case. Let S be $S_1 + S_2$, and A, A' be locally maximal physical quantities on S_1, S_2 respectively. (Note that from now on primed names of physical quantities refer to physical quantities for S_2.) Let $\hat{Q} = h(\hat{A})$ and $\hat{Q}' = k(\hat{A}')$ for (not necessarily locally maximal) physical quantities Q, Q' on S_1, S_2 respectively; and let $O(A, A')$ be a maximal physical quantity on $S_1 + S_2$ for which

$$\hat{O}(A, A') = \Sigma o_{ij} \hat{P}_i \times \hat{P}'_j, \text{ where } o_{ij} \neq o_{i'j'} \text{ for } i, j \neq i', j',$$

where the $\hat{P}_i$ are projectors onto the eigenvectors of A in $\mathscr{H}(S_1)$ and the $\hat{P}'_j$ are projectors onto the eigenvectors of A' in $\mathscr{H}(S_2)$ (i.e. $\hat{A}$

$= \Sigma a_i \hat{P}_i$ and $\hat{A}' = \Sigma a_j' \hat{P}_j'$). As such there exist f' such that

$$\hat{A} \times \hat{I} = f(\hat{O}(A, A')) \text{ and } \hat{I} \times \hat{A}' = f'(\hat{O}(A, A')),$$

(where the left-hand operator in a cross-product of form $\hat{X} \times \hat{Y}$ is taken to be an operator on $\mathscr{H}(S_1)$ and the right-hand operator an operator on $\mathscr{H}(S_2)$). Indeed we simply define f, f' by:

$$f(o_{ij}) = a_i \text{ and } f'(o_{ij}) = a_j \text{ for all } i, j.$$

Hence

(i) $\hat{Q} \times \hat{I} = h(\hat{A} \times \hat{I}) = h(f(\hat{O}(A, A')))$ and
       $\hat{I} \times \hat{Q}' = k(f'(\hat{O}(A, A')))$.

We can therefore apply (CVR) to $\hat{Q} \times \hat{I}$, $\hat{I} \times \hat{Q}'$ and $\hat{O}(A, A')$ in $S_1 + S_2$ (as substitution instances for $\hat{Q}, \hat{Q}', \hat{R}$, and S respectively), and hence

(ii)   If $\text{Prob}((Q \times I, I \times Q'), (q, q')) = 0$, where $\hat{Q}, \hat{Q}'$ are functions of locally maximal $\hat{A}, \hat{A}'$ respectively, then val $Q \times I_{|O(A, A')} \neq q$ or val $I \times Q'_{|O(A, A')} \neq q'$.

We also see that if S_1, S_2 are spatially separated then from OLOC it follows that

(iii) For any A', B' maximal on S_2 and any A maximal on S_1, val $A \times I_{|O(A, A')} = \text{val } A \times I_{|O(A, B')}$,

where $\hat{O}(A, B') = \Sigma o_{ij} \hat{P}_i \times \hat{Q}_j'$ the $\hat{Q}_j'$ being projectors onto the eigenvectors of B' in $\mathscr{H}(S_2)$.

But by (Func)* and (i)

$$\text{val } Q \times I_{|O(A, A')} = h(\text{val } A \times I_{|O(A, A')}).$$

Moreover $\hat{Q} \times \hat{I} = h(\hat{A} \times \hat{I}) = hg'(\hat{O}(A, B'))$ for some g'. So by (Func)*

$$\text{val } Q \times I_{|O(A, B')} = h(\text{val } A \times I_{|O(A, B')}).$$

Hence applying h to both sides of the equality (iii) gives the following weak form of OLOC:

AWOLOC   If S_1 and S_2 are spatially separated then for any A', B'
          maximal on S_2 and any A maximal on S_1, and for any $\hat{Q}$
          which is a function of $\hat{A}$, val $Q \times I_{|O(A, A')} = \text{val } Q$
          $\times I_{|O(A, B')}$.

And the same proof can be gone through with S_1 and S_2 interchanged to give:

BWOLOC   If S_1 and S_2 are spatially separated then for any A, B maximal on S_1 and any A' maximal on S_2, and for any $\hat{Q}'$ which is a function of $\hat{A}'$, $\text{val } I \times Q'_{|O(A, A')} = \text{val } I \times Q'_{|O(B, A')}$.

We now specialize further by assuming not only that S_1 and S_2 are spatially separated but that $S_1 + S_2$ at t is in the state $F = \Sigma c_i f_i \times f'_i$ where f_i, f'_i are the eigenvectors in $\mathscr{H}(S_1)$ and $\mathscr{H}(S_2)$ of the locally maximal A, A' respectively. And we suppose that B is a locally maximal physical quantity for S_1 with Q another physical quantity for S_1 for which $\hat{Q} = h(\hat{A}) = k(\hat{B})$, e.g. $\hat{Q}$ may be $\hat{P}(f_1) = \hat{P}(g_1)$, where f_1 and g_1 are eigenvectors of B and A respectively. Note $\hat{Q}, \hat{B}$, and $\hat{A}$ are here taken as operators on $\mathscr{H}(S_1)$.

Now from the Born interpretation, because of the diagonal form of F, we derive:

Prob $((A \times I, I \times A'), (a_i, a')) = 0$ for $a' \neq a'_i$.

Hence, from (ii),

If val $(A \times I)_{|O(A, A')} = a_i$ then val $(I \times A')_{|O(A, A')} \neq a'$ for any $a' \neq a'_i$, i.e.

(iv)   If val $(A \times I)_{|O(A, A')} = a_i$ then val $(I \times A')_{|O(A, A')} = a'_i$.

But also

Prob $((A \times I, I \times A'), (a, a'_i)) = 0$ for any $a \neq a_i$;

and hence, since from the Born interpretation it follows that

$$\text{Prob}((h(A \times I), I \times A'), (h(a), a'_i)) = \sum_{j : h(a_j) = h(a)} \text{Prob}((A \times I, I \times A'), (a_j, a'_i)),^{10}$$

we see that

Prob $((h(A \times I), (I \times A')), (h(a), a'_i)) = 0$ for $h(a) \neq h(a_i)$.

Hence, since $\hat{Q} \times \hat{I} = h(\hat{A} \times \hat{I})$,

Prob $((Q \times I, I \times A'), (h(a), a'_i)) = 0$ for $h(a) \neq h(a_i)$.

---

[10] I.e. the left-hand side $= \text{Tr } \hat{W} \sum\limits_{j : h(a_j) = h(a)} \hat{P}_j \times \hat{P}'_i = \sum\limits_{j : h(a_j) = h(a)} \text{Tr } \hat{W} \hat{P}_j \times \hat{P}'_i$,

And hence, since $\hat{Q} \times \hat{I}$ is also a function of $(\hat{B} \times \hat{I})$, by (ii),

(v)   If val $I \times A'_{|O(B, A')} = a'_i$, then val $Q \times I_{|O(B, A')}$
$$\neq h(a) \text{ for any } h(a) \neq h(a_i), \text{ and}$$

so val $Q \times I_{|O(B, A')} = h(a_i)$.

But from BWOLOC it follows that

val $I \times A'_{|O(B, A')} = $ val $I \times A'_{|O(A, A')}$,

and hence substituting into (v) gives:

(vi)   If val $I \times A'_{|O(A, A')} = a'_i$, then val $Q \times I_{|O(B, A')} = h(a_i)$.

From (iv) and (vi) it then follows that

(Func R)   If $\hat{Q} = h(\hat{A})$ and val $A \times I_{|O(A, A')} = a_i$, then val $Q \times I_{|O(B, A')}$
$= h(a_i)$ for any locally maximal B, A, A' for which $\hat{Q} = k(\hat{B})$
for some k.

This last principle (Func R) is the main result. It is not the same as
(Func), but is similar enough to (Func) in its restriction on functional
relationships between the values of physical quantities that a
contradiction can be derived from it via the KS argument. HR show
this for the case where S_1, S_2 are spin 1 systems and F is the singlet
state. And, in terms of our completeness assumption (C1), we can see
how this contradiction is generated as follows.

Consider a particular f in $\mathcal{H}(S_1)$ which is the eigenvector of both
the locally maximal A, B for eigenvalues a_i and b_j respectively.
Consider what in the terms of section 1 we would denote 'A_i', i.e. by
definition

val $A_i = 1$ if val $A = a_i$

    „ $= 0$ „ $= a_j$ for any $j \neq i$,

and similarly define B_j.

Then by (C1) we have $\hat{A}_i = h(\hat{A})$ and $= k(\hat{B})$ for some h, k. And hence,
by (Func R),

(vii)   val $A_i \times I_{|O(B, A')} = 1$ if val $A \times I_{|O(A, A')} = a_i$

        „ $= 0$ „ $= a_j$ for any $j \neq i$.

But

val $B_j \times I_{|O(B, A')} = $ val $A_i \times I_{|O(B, A')}$, since *ex hypothesi* $\hat{B}_j = \hat{A}_i$. And since, similarly to (vii), we can derive:

(viii) val $B_j \times I_{\mid O(B,\,A')} = 1$ if val $B \times I_{\mid O(B,\,A')} = b_j$

     " $= 0$ " $= b_k$ for any $k \neq j$,

it follows that

(ix) val $A_i \times I_{\mid O(B,\,A')} = 1$ if val $B \times I_{\mid O(B,\,A')} = b_j$

     " $= 0$ " $= b_k$ for any $k \neq j$.

Hence from (vii) and (ix) it follows that

 val $A \times I_{\mid O(A,\,A')} = a_i$ iff val $B \times I_{\mid O(B,\,A')} = b_j$.

Hence we can associate with each f a unique number $m(f)$—be it 1 or 0—which is val $A \times I_{\mid O(A,\,A')}$ for any A for which f is an eigenvector of A in $\mathcal{H}(S_1)$. The m so defined is easily seen to satisfy the conditions that it is single valued, $m(f) = 1$ or 0, and $\Sigma\, m(f_i) = 1$, generating the KS contradiction. Thus as promised a contradiction has been derived from (VR), (OLOC), (Det Q), and Standard QT. In other words a realist interpretation of QT, which includes (Det Q), (Pass Q)', (C1), and (VR)', and hence (VR) too, must violate OLOC and more particularly BWOLOC.

3. OLOC

In this section we shall discuss the 'locality principle', OLOC. At first sight it seems that we can show that OLOC, in the weak form BWOLOC in which it partakes in the previous proof, is a locality condition if we accept conservative de-Ockhamization and in particular the (Pass Q)* of section 1. The argument for this is the following:

 Consider a failure of BWOLOC, i.e. suppose that

 val $I \times A'_{\mid O(B,\,A')} \neq$ val $I \times A'_{\mid O(A,\,A')}$.

Now consider a measurement of $O(B, A')$. Since $O(B, A')$ is maximal on $S_1 + S_2$ we can take all ideal methods of measuring it on $S_1 + S_2$ to agree (given (Pass Q)*). And one such way to measure it is by a joint measurement of B and A'. (This is because if $\hat{B} = \Sigma b_i \hat{P}_i$ and $\hat{A}' = \Sigma a'_j \hat{P}'_j$ then the conditions specified by QMT for a joint measurement of B and A' satisfy the conditions specified by QMT for a measurement of a physical quantity with representative operator $\hat{O}(B, A') = \Sigma o_{ij} \hat{P}_i \times \hat{P}'_j$.) Moreover for such a measurement QMT implies that the measured value of $O(B, A')$ is o_{ij} iff the corresponding joint measured values of B and A' are b_i and a'_j respectively. Let the measured value of $\hat{O}(B, A')$ obtained in this way be $\mathrm{mv_{BA'}}O(B, A')$. Thus we have, by (Pass Q)*,

$$\mathrm{mv_{BA'}}O(B, A') = \mathrm{val}\ O(B, A').$$

But by (Func)*,

$$\mathrm{val}\ I \times A'_{\mid O(B, A')} = f(\mathrm{val}\ O(B, A')),\ \text{where}\ f\ \text{is given by}$$
$$\hat{I} \times \hat{A}' = f(\hat{O}(B, A')),\ \text{i.e.}\ f(o_{ij}) = a'_j\ \text{for all}\ i, j.$$

Hence

$$\mathrm{val}\ I \times A'_{\mid O(B, A')} = f(\mathrm{mv_{BA'}}O(B, A')).$$

Now let $\mathrm{mv_{BA'}}O(B, A') = o_{ij}$, so that $\mathrm{mv_{BA'}}A' = a'_j$ and $\mathrm{mv_{BA'}}B = b_i$.

Hence

$$f(\mathrm{mv_{BA'}}O(B, A')) = a'_j = \mathrm{mv_{BA'}}A'.$$

And hence

$$\mathrm{val}\ I \times A'_{\mid O(B, A')} = \mathrm{mv_{BA'}}A'.$$

And similarly we derive:

$$\mathrm{val}\ I \times A'_{\mid O(A, A')} = \mathrm{mv_{AA'}}A'.$$

Hence the failure of BWOLOC entails

$$\mathrm{mv_{AA'}}A' \neq \mathrm{mv_{BA'}}A',$$

which clearly does constitute a locality failure, since it means that A', a physical quantity for S_2, has a measured value which depends on what is commeasured with it on the spatially separated S_1. QED.

So it seems that for a conservative realist interpretation of QT (one which incorporates a conservative de-Ockhamization strategy including (Pass Q)*) BWOLOC is indeed a locality condition in the sense that a BWOLOC failure means locality failure; and hence a conservative realist interpretation of QT, which in the previous section we showed must violate BWOLOC, must also violate locality. (Although as indicated HR would not accept *conservative* de-Ockhamization, and hence BWOLOC does not have this connection with locality for them.)

I shall argue, however, that this BWOLOC is a consequence of QT and a principle which is central to the notion of a physical quantity in both the classical and quantum theoretic context ('analytic' one may even say). This means that the premisses of the immediately preceding argument, which combined a conservative realist interpretation of QT with rejecting BWOLOC and hence locality, are not consistent. More importantly, however, it means that the arguments of the previous sections can be taken immediately to show that realism, *qua* the principles (Det Q) and (Pass Q)', contradict a QT which has been standardly de-Ockhamized (and hence includes (C1)). This is because from section 1 we see that QT, including (VR)' plus (C1) and (Pass Q)', imply (VR), and from section 2 we then see that (VR) plus QT and (Det Q) contradict BWOLOC. In short it is not just the disjunction realism *or* locality which is inconsistent with QT (as Bell *et al.* claim) but it is realism alone which is inconsistent with QT.

Consider the range of values which the various physical quantities for S_1 which are represented by $\hat{A}$ on $\mathscr{H}(S_1)$ have at a given time t. In the case that $\hat{A}$ is maximal on $\mathscr{H}(S_1)$ this range of values consists of just the one value given the usual de-Ockhamization discussed in section 1, i.e. we have

(U)*   If A, A' are physical quantities for S_1 which are represented by the same maximal $\hat{A}$ on $\mathscr{H}(S_1)$ then the value of A at t is equal to the value of A' at t. Indeed $A = A'$ intensionally.

(We will not be needing the final sentence of (U)* in what follows. I shall refer to the version of (U)* which includes this final sentence as Strong (U)*.) But it can be taken as a principle of QT (I defend this below) that

$(P)_0$ A is a physical quantity for S_1 represented by $\hat{A}$ on $\mathscr{H}(S_1)$ if it is also a physical quantity for $S_1 + S_2$ represented by $\hat{A} \times \hat{I}$ on $\mathscr{H}(S_1 + S_2)$.

From (U)* and $(P)_0$ it follows that all physical quantities for $S_1 + S_2$ which are represented by $\hat{A} \times \hat{I}$ on $\mathscr{H}(S_1 + S_2)$, where $\hat{A}$ is maximal on $\mathscr{H}(S_1)$, have the same value at t (indeed they are all the same physical quantity according to Strong (U)*). Hence

If $\hat{A}$ is maximal on $\mathscr{H}(S_1)$ then $\text{val}_{S_1 + S_2} A \times I_{|X} = \text{val}_{S_1 + S_2} A \times I_{|Y}$ for any X, Y which are maximal on $S_1 + S_2$,

which immediately gives us OLOC and hence BWOLOC. QED.

Note 1 The number of physical quantities which are represented by a non-maximal operator $\hat{O}$ on $\mathscr{H}(S_1 + S_2)$ may be more than one, and they may all have different values, once we allow de-Ockhamization. But we have just shown that, in the special case that $\hat{O}$ takes the form $\hat{A} \times \hat{I}$ for maximal $\hat{A}$ on $\mathscr{H}(S_1)$, even though $\hat{A} \times \hat{I}$ is non-maximal the various physical quantities represented by $\hat{A} \times \hat{I}$ must all have the same value. Moreover, although we can still formally distinguish the various physical quantities $A \times I_{|X}$ and $A \times I_{|Y}$, $(P)_0$ together with the strong version of $(U)^*$ imply not only that these various physical quantities have the same value but also that they are the same physical quantity.

Note 2. This derivation of OLOC does not require S_1 and S_2 to be spatially separate (although it does require them to be distinct and in particular to have distinct Hilbert spaces), and hence an even stronger version of OLOC has been derived than was stated in the text above.

Note 3. There is an implicit premiss in the previous argument which allows us to refer to 'the value of A at t' without specifying a system in the context of which A is being considered. This premiss can be put formally as the proposition that

$(P)_1$ If A is a physical quantity for both S and S' then $A(S, t) = A(S', t)$.

It is this premiss which allows us to infer from $A(S_1, t) = A'(S_1, t)$ to $A(S_1 + S_2, t) = A'(S_1 + S_2, t)$. I take this last premiss $(P)_1$ to be effectively analytic, a presupposition of our discourse (in both QT and classical mechanics) about physical quantities in sub-systems. Thus the required result has been achieved: OLOC derived from QT and effectively analytic principles.

One point at which the preceding argument can be pressed is on its assumption $(P)_0$. In particular it may be denied that $(P)_0$ really is a law of QT. Rather than embark on a verbal dispute over which propositions really belong to QT I shall simply present an argument for $(P)_0$: Under what conditions will the physical quantity A with possible values $\{a_i\}$ be counted as a q-quantity for S_1? Surely one way is for there to be a complete orthonormal set of vectors $\{f_i^1,\}$ in $\mathscr{H}(S_1)$ for which, for any f^1 in $\mathscr{H}(S_1)$, the probability of measuring A (however we do that) to have value a_i at t is $|(f_i^1, f^1)|^2$ if $f(S_1, t) = f^1$. Indeed if these conditions are satisfied then A is surely not only a q-quantity for S_1 but is represented by the operator $\Sigma\, a_i \hat{P}(f_i^1)$ on S_1. But clearly it is exactly these conditions which are satisfied if we suppose that A is represented by $\hat{A} \times \hat{I}$ on $\mathscr{H}(S_1 + S_2)$ for some maximal $\hat{A}$

$= \Sigma a_i \hat{P}(f_i^1)$. I.e. if we let $f(S_1, t) = f^1$ then the most general form for $\hat{W}(S_1 + S_2, t)$ is $\hat{P}(f^1) \times \hat{W}$ for some density operator $\hat{W}$ on $\mathcal{H}(S_2)$.[11] But since (*ex hypothesi*) A is represented by $\hat{A} \times \hat{I}$ on $\mathcal{H}(S_1 + S_2)$, the projection operator onto the subspace of its eigenvectors in $\mathcal{H}(S_1 + S_2)$ for possible value a_i must be $\hat{P}(f_i) \times \hat{I}$.[12] (BSI)' then gives the required conditions. QED.

CONCLUSION

OLOC is, I have argued in section 3, a consequence of (standardly de-Ockhamized) QT (including $(P)_0$) together with a certain analytic principle about physical quantities in sub-systems (viz. $(P)_1$). And (VR), I have argued in section 1, is a consequence of adopting a de-Ockhamization which is conservative enough to both include (Func)* and be minimally realist in respect of assuming (Pass Q)' (although, as indicated, some Copenhageners would reject this). But from OLOC and (VR) we can derive a contradiction with the realist principle (Det Q) within the standard framework of QT. (I showed this in section 2 following a proof by HR.) Thus a realist interpretation of QT seems hopeless unless the standard framework of QT is weakened in some way or we opt for a more radical de-Ockhamization (which gives up (Func)* for example). And indeed one way to weaken QT so that the above proofs break down is to weaken (VR) to (VRR) and consequently weaken the eigenvector rule, e.g. by contextualizing it to

(ERR) If $f(S, t)$ is an eigenvector of Q for eigenvalue q, then
$Q(S, t)(Q) = q$.

However the HR proof shows that even if we make this modification of QT then there is still a contradiction between realism and QT if we additionally assume another locality condition ELOC.[13] In short *ad hoc* shifts within the formalism of QT, such as weakening (ER), do not rescue realism in any satisfactory way. The lesson seems clear that the inconsistency between QT and the traditional realist framework

---

[11] See Appendix 7.

[12] i.e. for any f_i^1, g, $\hat{A} \times \hat{I} f_i^1 \times g = a_i f_i^1 \times g$; and hence the eigenvectors of A in $\mathcal{H}(S_1 + S_2)$ for possible value a_i consist of the closed linear subspace of all vectors of form $f_i^1 \times g$.

[13] The HR proof is summarized in Appendix 12. Note that HR points out (1983) p. 498 that ELOC is a locality condition only in the situation where we assume OLOC, as we do here.

which I have demonstrated here is not to be removed by *ad hoc* adjustments to QT. To accept QT it seems we must reject realism, and not just the combination of locality and realism. Realism cannot be rescued even by going 'non-local'.

Moreover since the modified Bohr interpretation which I have been arguing for in this book takes all *q*-quantities to have determinate values in the context of the states considered in the Heywood–Redhead argument, it must be the realist principle (Pass Q) in one of its forms which I am committed to giving up (rather than just (Det Q)). In particular it seems that I must give up the weak form of the passivity expressed in (VR), and more especially in (Pass Q)'.

However, let me repeat the qualifications made earlier in this book. The term 'realism' as it appears in this last conclusion has a meaning which is quite idiosyncratic to discourse on the foundations of QT. To give up realism, as I am advocating here, is only to admit the impossibility of grafting certain groups of classical principles, in particular (Pass Q) in some of its various forms, onto QT. The issue of whether realism in some more usual sense(s) must be surrendered in the light of QT is a quite different issue. And in this book I have argued steadfastly in favour of realism in its more usual senses. Thus contrary to both Bohr and Cartwright, I have argued for an entirely realist, as opposed to instrumentalist, interpretation of QT: one which takes seriously the formalism of QT as describing the world in some reasonably one-to-one fashion. Moreover I have also supported a realist, *qua* objectivist, metaphysics for QT, by opposing those Copenhagen views which see the modes of description for *q*-systems as 'subjective' in the sense that they depend on what we choose to measure. And finally I have opposed those subjectivist metaphysics for QT which interpret various of the entities referred to in QT—the probabilities, the indeterminacies—as subjective. In short, realism in the specialized sense of (Det Q) and (Pass Q) may have to be given up, but QT provides no reason for rejecting realism in any broader sense.

Operators and Physical Quantities

In order to describe the representation of discrete valued q-quantities in a way which will later allow us to generalize to continuous valued q-quantities, we shall need to introduce the notion of an *operator* on a Hilbert space in more detail than in the main text. An operator in the Hilbert space $\mathscr{H}$ is defined as an operation which transforms vectors in $\mathscr{H}$ into other vectors in $\mathscr{H}$. We will denote such operators by $\hat{E}, \hat{F}, \hat{G}. \ldots, \hat{P}, \hat{Q}, \hat{R} \ldots$; and let '$\hat{E}g$' denote the vector into which the vector g in $\mathscr{H}$ is transformed by the operator $\hat{E}$.

All the operators we shall be considering will be 'linear'. To explain more fully what this means we must first introduce the idea of the domain of an operator:

> The domain of the operator $\hat{E}$ on $\mathscr{H}$ is that subset of the vectors f in $\mathscr{H}$ for which $\hat{E}f$ is a vector in $\mathscr{H}$.

(Note that some operators in QT cannot have a domain equal to the whole of $\mathscr{H}$. This, as Toeplitz's theorem shows[1], follows essentially from their unboundedness, and this in turn follows from the fact that they represent physical quantities the possible values of which are unbounded in magnitude.) We can now define the notion of linearity:

Defn 1 $\hat{E}$ is linear iff not only is its domain a linear subspace (i.e. if f_1, $f_2, \ldots f_N$ belong to the domain, so does $(c_1 f_1 + c_2 f_2 + \ldots c_N f_N)$, for any real numbers $\{c_i\}$) but also $\hat{E}(c_1 f_1 + c_2 f_2 + \ldots + c_N f_N) = c_1 \hat{E} f_1 + c_2 \hat{E} f_2 + \ldots c_N \hat{E} f_N$.

All the operators we will be considering will also be 'self-adjoint' or 'hermitean' in the following sense:

Defn 2 $\hat{E}$ is self-adjoint iff for any f, g in the domain of $\hat{E}$, $(\hat{E}f, g) = (f, \hat{E}g)$;

and they will also all be 'closed' in the following sense:

Defn 3   An operator is closed iff not only is its domain a closed linear subspace, i.e. if $\{f_i\}$ is a convergent sequence of vectors in the domain then the limit of the sequence is also in the domain, but also

---

[1] Von Neumann (1955), p. 150.

if the limit of f_i as i tends to infinity is f then $\hat{E}f_i$ tends to $\hat{E}f$ as i tends to infinity.

From this it follows that

If $\Sigma\, c_i f_i$ exists in $\mathscr{H}$, where $\{f_i\}$ are all in the domain of $\hat{E}$, then $\hat{E}\Sigma c_i f_i$ $= \Sigma\, c_i \hat{E}f_i$.

In other words linearity extends to infinite sums.

The notion of a 'projection operator', which has so far been defined only for a physical vector space, can be extended to apply to abstract vector spaces. These abstract projection operators turn out to be of central importance in QT.

Defn 4 $\hat{P}(f)$, the projection operator onto the vector f in $\mathscr{H}$, is the operator for which $\hat{P}(f)g = (g,f)f$, for any g in $\mathscr{H}$.

Defn 5 $\hat{P}(V)$, the projection operator onto the closed linear subspace V in $\mathscr{H}$, is the operator for which for any set $\{f_i\}$ which is a basis set for V, and for any g in $\mathscr{H}$, $\hat{P}(V)g = \Sigma\, \hat{P}(f_i)g$.

To show this definition is consistent it must of course be proven that $\Sigma\, \hat{P}(f_i)g = \Sigma\, \hat{P}(f_i')g$ for any g in $\mathscr{H}$, and any two basis sets $\{f_i\}$ *and* $\{f_i'\}$ for $\mathscr{H}$. This is easily done q.v.:

Theorem   Let $\{f_i\}$ and $\{f_i'\}$ both be basis sets for some closed linear subspace V. Then $\Sigma\, \hat{P}(f_i) = \Sigma\, \hat{P}(f_i')$.

Proof   Let $\{g_j\}$ be an orthonormal set such that $\{f_i\} \cup \{g_j\}$ is a basis set for all of $\mathscr{H}$. Then for any g in $\mathscr{H}$

$$g = \Sigma\, c_i f_i + \Sigma\, d_j' g_j.$$

But $\{f_i'\} \cup \{g_j\}$ is clearly a basis set for all of $\mathscr{H}$ too, i.e. this set is obviously orthonormal, and moreover it is easily seen to be complete. (It is complete because $\{f_i\} \cup \{g_j\}$ is complete, and because any vector in that set is a linear combination of the vectors in $\{f_i'\} \cup \{g_j\}$.) Hence

$$g = \Sigma\, c_i' f_i' + \Sigma\, d_j g_j.$$

But

$$d_j = d_j' = (g_j, g).$$

Hence

$$\Sigma\, c_i' f_i' = \Sigma\, c_i f_i = g - \Sigma\, d_j g_j.$$

But clearly

$$\Sigma\, \hat{P}(f_i)g = \Sigma\, c_i f_i$$
$$\Sigma\, \hat{P}(f_i')g = \Sigma\, c_i' f_i'.$$

Hence

$$\Sigma\, \hat{P}(f_i)g = \Sigma\, \hat{P}(f_i')g,$$ and the required result follows. QED.

Corollary If g belongs to V then $\Sigma \hat{P}(f_i)g = g$.

Proof If g belongs to V then $g = \Sigma c_i f_i$ (since $\{f_i\}$ is a basis set for V). Trivially the result then follows. QED.

Also we can define, in the obvious way, operations of addition, scalar multiplication and product for operators on $\mathcal{H}$:

Defn 6 $(c_1\hat{E} + c_2\hat{F})$ is that operator on $\mathcal{H}$ for which, for any g in the domain of both $\hat{E}$ and $\hat{F}$, $(c_1\hat{E} + c_2\hat{F})g = c_1\hat{E}g + c_2\hat{F}g$ (and which has a domain containing only such g).

Defn 7 $(\hat{E}\hat{F})$, the product of $\hat{E}$ with $\hat{F}$, is that operator on $\mathcal{H}$ for which, for any g in the domain of $\hat{F}$ for which $\hat{F}g$ is in the domain of $\hat{E}$, $(\hat{E}\hat{F})g = \hat{E}(\hat{F}g)$ (and which has a domain containing only such g).

From Defns. 5 and 6 we then trivially see that

$$\hat{P}(V) = \Sigma \hat{P}(f_i) \text{ for any } \{f_i\} \text{ which is a basis set for } V.$$

It is easily seen that projection operators are 'idempotent', i.e. $\hat{P}(V)\hat{P}(V) = \hat{P}(V)$. This is because if we project a vector onto a subspace V, and then project the result of doing that onto the subspace V again, then the second projection produces no further change.

We can define infinite sums of operators q.v. . Let $\Sigma c_i\hat{Q}_i$ be that operator which transforms the vector g in $\mathcal{H}$ into the limit as N tends to infinity of $\sum_{i=1}^{N} c_i(\hat{Q}_i g)$ if such a limit exists in $\mathcal{H}$. If the limit does not exist in $\mathcal{H}$ then g is said not to be in the domain of $\mathcal{H}$. (If the sum has a finite number of terms then clearly $\Sigma c_i\hat{Q}_i$ has the whole of $\mathcal{H}$ as its domain if each of the $\hat{Q}_i$ does; but if the sum has infinitely many terms then this may not be the case). It is easily shown that $\Sigma c_i\hat{Q}_i$ so defined is still linear, closed and hermitean if the various $\hat{Q}_i$ are.

What is the point of introducing these operators on a Hilbert space? It is because it turns out to be convenient to associate with every physical quantity Q (for system S) a linear hermitean closed operator $\hat{Q}$ (on $\mathcal{H}(S)$) which has the following defining feature:

(i) For any possible value q of Q, $\hat{Q}f = qf_i$ for any and only those f which are eigenvectors of Q for the possible value q,

where (see Chapter 2) f is defined to be an 'eigenvector' of the physical quantity Q for possible value q iff f is a possible state for which if S at t is in that state then Q has the value q in S at t with certainty. In defining eigenvectors in this way we have reversed the usual order of things, which is to start from the idea of representing a physical quantity Q by an operator $\hat{Q}$, and then define f to be an eigenvector of $\hat{Q}$ for eigenvalue q iff (i) is true. The reason for making this reversal is to avoid using the sophisticated concept of an operator until the last possible minute. We are now finally bringing these two different concepts of an eigenvector together.

If we assume that the set of eigenvectors for a q-quantity includes a basis set then it follows that (i), together with the condition that $\hat{Q}$ is closed, linear and hermitean, is sufficient to uniquely determine $\hat{Q}$, i.e. for any basis set $\{f_i\}$ any vector g can be written as $\Sigma c_i f_i$ for some choice of $\{c_i\}$, and hence

$$\hat{Q}g = \hat{Q}\Sigma c_i f_i,$$

which because of the linearity of $\hat{Q}$ (and its closure) means that

$$\hat{Q}g = \Sigma c_i \hat{Q} f_i.$$

But now let $\{f_i\}$ be a set of eigenvectors of Q, with f_i the eigenvector of Q for possible value q_i. (Since Q need not be non-degenerate we may have $q_i \neq q_j$ for $i \neq j$.) Then, by (i),

$$\hat{Q}g = \Sigma c_i q_i f_i.$$

Thus $\hat{Q}g$ is uniquely determined for any g if it exists, i.e. if $\Sigma c_i q_i f_i$ is convergent, which it may not be if $\hat{Q}$ is unbounded, as indicated above. Thus $\hat{Q}$ is uniquely determined since to uniquely determine $\hat{Q}$ is to uniquely determine $\hat{Q}g$ for any g in the domain of $\hat{Q}$ (the domain of $\hat{Q}$ being simply the set of g for which $\hat{Q}g$ exists).

There is still of course the question of whether such $\hat{Q}$ exist which satisfy (i) and are closed, linear, and hermitean. But this too is answered in the affirmative at least for discrete-valued Q if we make the following additional assumptions, all of which are made in QT:

(Ass) For discrete-valued Q the set of eigenvectors for Q includes a basis set $\{f_i\}$ for which the corresponding set of possible values $\{q_i\}$ exhausts all the possible values for Q. Moreover the eigenvectors of Q for possible value q_i form a closed linear subspace $V(Q, q_i)$ such that those of the $\{f_i\}$ which belong to $V(Q, q_i)$ provide a basis set for $V(Q, q_i)$.

We now prove that under these assumptions some $\hat{Q}$ do indeed exist satisfying the conditions just given:

Proof Let Q be discrete-valued. Let $\hat{Q}^* = \Sigma q_i \hat{P}(f_i)$, where $\{f_i\}$ is any basis set of eigenvectors of Q. We now show that the set of vectors f for which $\hat{Q}^* f = qf$ coincides with the set of eigenvectors of Q for the possible value q, for any such q. The proof comes in two parts.

First, let f be any eigenvector of Q for possible value q. Then, by (Ass), f is in a closed linear subspace for which a basis set is formed by those members of $\{f_i\}$ which are eigenvectors of Q for possible value q. Let those eigenvectors be $\{f_1, f_2, \ldots, f_n\}$. Then, by the corollary to the theorem following Defn. 5,

$$\sum_1^n \hat{P}(f_i)f = f$$

and of course we also have

$q_i = q$ for all $i \leq n$.

Now consider f_j for which $j > n$. Then since $f = \sum_1^n c_i f_i$, and since (f_i, f_j) $= 0$ for $i \leq n$, we clearly have

$$(f_j, f) = \sum_{i=1}^n c_i(f_j, f_i) = 0.$$

Hence $\hat{P}(f_j)f = 0$ for any f_j for which $j > n$.

Thus we see that

$$\hat{Q}^* f = \sum q_i \hat{P}(f_i) f = \sum_1^n q_i \hat{P}(f_i) f + \sum_{i>n} q_i \hat{P}(f_i) f$$

$$= \sum_1^n q \hat{P}(f_i) f + 0$$

$$= q \sum_1^n \hat{P}(f_i) f = qf.$$

Thus we have shown that any eigenvector of Q for possible value q satisfies $\hat{Q}^* f = qf$.

Now the second part of the proof.

Let $\hat{Q}^* f = qf$, i.e.

$$\sum q_i(f_i, f) f_i = q \sum (f_i, f) f_i.$$

Hence, since $\{f_i\}$ is a basis set, we have

$q_i(f_i, f) = q(f_i, f)$ for all i,

i.e. for any i *either* $q_i = q$ (if $(f_i, f) \neq 0$) *or* $(f_i, f) = 0$. But $(f_i, f) \neq 0$ for some i (since $\{f_i\}$ is complete). Hence $q_i = q$ for some i. Moreover $(f_i, f) = 0$ for any i for which $q_i \neq q$. This last condition (together with the completeness of the $\{f_i\}$) is then easily seen to imply that f is in $V(Q, q)$.

From these two parts of the proof together we see that, $\hat{Q}^*$, which is clearly closed, hermitean, and linear, satisfies (i). Hence there are closed, linear, hermitean operators satisfying (i) for Q discrete-valued. QED.

Thus we have shown that if Q is discrete-valued then there is a unique closed linear hermitean operator $\hat{Q}$ associated with it satisfying (i). This operator may be said to represent Q in that it summarizes all the information which we have taken to characterize a q-quantity, viz. its eigenvectors and possible values. It summarizes this information via the equation (i), i.e. the set of $\{q, f\}$ which satisfy (i) are the eigenvectors and possible values (or eigenvalues) of Q.

Advantages of the Operator Representation

There are two main advantages of the operator representation for physical quantities (as opposed to representation in terms of eigenvectors and possible values which I have been using hitherto for ease of exposition). To appreciate the first reason it must be realized that the transition from classical mechanics CM to QT can be characterized as a change in the way physical quantities are represented, albeit a change in which the notion of an algebraic structure for physical quantities is preserved.

In CM physical quantities are represented by real-valued functions while in QT they are represented by operators on a Hilbert space. Thus in CM the energy of some system S is represented by a function which maps times onto real numbers, viz. onto possible energy values. In QT on the other hand the energy of a system S is represented by an operator for which (at least in the case where energy is discrete-valued) the eigenvalues are the possible energy values, and the eigenvectors are those states for which when S is in one of those states it has a particular energy value with certainty.

The operators representing physical quantities in QT have a certain algebraic structure: they can be added, multiplied by a constant, and form products (see Defns. 6, 7, Appendix 1). But the functions representing physical quantities in CM also have an algebraic structure: they can be added, multiplied by a constant, and form products (i.e. be multiplied together.) The product of the function f with the function g—denoted '$f \cdot g$'—is defined as the function satisfying $(f \cdot g)(x) = f(x)g(x)$.

This 'product' of the two functions f and g must be distinguished from their 'composition' which is denoted '$f(g)$', and defined q.v.: $f(g)$ has the value $f(g(x))$ for argument x iff x is in the domain of g and $g(x)$ is in the domain of f. Note that it is the composition of two functions, rather than their product (as just defined), which is defined in an analogous fashion to the product of the two operators—see Defn 7 in Appendix 1.

But, and here is the crucial point, the algebraic structures of functions and operators are importantly different. Whereas the algebra of functions is a commutative algebra, the algebra of operators is not, i.e. although

$$f \cdot g = g \cdot f,$$

which means that for any x which is in the domain of both the functions f, g, $f(x)g(x)=g(x)f(x)$, we nevertheless do not always have

$$\hat{E}\hat{F} = \hat{F}\hat{E},$$

i.e. the transformation of a vector by the operation $\hat{F}$, followed by the operation $\hat{E}$, does not always have the same result as the transformation of the same vector by $\hat{E}$, followed by $\hat{F}$. (For example to rotate a vector f by 90° about an axis at right-angles to f, and then to rotate the result by 90° about an axis along f clearly has a different result from performing the rotations in the reverse order.)

This disanalogy between the algebraic structures arises because we have not taken the composition of two functions as their product. Had we insisted on letting their product be their composition (and to do so would have been to define the product of two functions in a fashion much more analogous to the product of two operators) then clearly the product operation on functions would not have been commutative either, since it is clearly possible that $f(g(x))\neq g(f(x))$, e.g. twice the square of a number is generally different from the square of twice the number.

So why do we seek to compare the algebraic structure of sums/products of functions with the algebraic structure of sums/products of operators, after defining the product of operators in this peculiar way? The reason is simply that if we define operator products in this way then there is a remarkable, if limited, correspondence between those functions (and their sums/products) which represent physical quantities in CM, and those operators (and their sums/products) which represent the same physical quantities in QT. This correspondence has been of great heuristic significance in building up QT, and here lies the first of the advantages of an operator representation. The correspondence in question is this:

(Corresp)   Let Q_1, Q_2 be two physical quantities. Then the physical quantity corresponding to their classical product/sum is represented in QT by an operator which is the product/sum of $\hat{Q}_1, \hat{Q}_2$ (although the product/sum is often only taken to be significant if $\hat{Q}_1, \hat{Q}_2$ commute).

In other words the physical quantity which would classically be represented by the function $(Q_1)+(Q_2)$, where (Q_i) is the function which represents Q_i in CM, is represented by the operator $\hat{Q}_1 + \hat{Q}_2$ in QT, and the physical quantity which would classically be represented by $(Q_1)(Q_2)$ (this is a product not a composition) is represented by $\hat{Q}_1\hat{Q}_2$ in QT. This correspondence is remarkable given the difference between the two concepts of product employed for functions and operators. (To appreciate the difference realize that (Corresp) is false if we replace the product of two operators by the composition of two functions throughout even if $\hat{Q}_1, \hat{Q}_2$ are assumed to commute, i.e. it is false that if $f(g(x))=g(f(x))$ for any x then $f(x)g(x)=g(x)f(x)$. To see this let $f(x)=1+x$ and $g(x)=2+x$.)

This correspondence means that starting from just a few operators for certain basic physical quantities we can build up the operators for a whole range of other physical quantities, viz. the sums, products (and of course scalar multiples) of the basic quantities. For example the classical relation: $E=(p^2/2m)+V$ (where E is the total energy, $p^2/2m$ is the kinetic energy, p being the momentum, m the mass, and V the potential energy) is transformed in QT into the Schrödinger equation:

$$i\hbar d/dt = -(\hbar^2/2m)d^2/dx^2 + V,$$

where $i\hbar d/dt$ is the operator representing energy, $i\hbar d/dx$ is the operator representing momentum, and V the operator representing potential energy. Note that in this case, the correspondence applies despite the non-commutativity of the operators concerned.

The second reason for using an operator representation for physical quantities becomes apparent in Appendix 4. It is that in the step up to considering continuous-valued physical quantities the representation in terms of operators is preserved, but not the representation in terms of eigenvectors and possible values.

APPENDIX 3

Degenerate Physical Quantities

In addition to the non-degenerate q-quantities which we have dealt with so far (for which there is a unique eigenvector for each possible value) there are degenerate q-quantities. Q' is said to be degenerate iff there are several different eigenvectors for at least one of its possible values. We can derive an expression for $P(Q', q_i', S, t)$, for Q' degenerate by making the following assumptions. Let $f_{i1}, f_{i2}, \ldots$ be some maximal mutually orthogonal subset of the eigenvectors for Q' for possible value q_i'. There will be some such subset for each value of i. We assume that there is a non-degenerate q-quantity Q which for all i has $f_{i1}, f_{i2}, \ldots$ as its eigenvectors for possible values $q_{i1}, q_{i2}, \ldots$ respectively. We then assume that a Q-measurement is also a Q'-measurement, and that whenever Q is measured to have value q_{in} then Q' is measured to have value q_i' (this is assumption (A) of p. 98); and so, for all i,

(i) $\quad P(Q', q_i', S, t) \geq \sum_n P(Q, q_{in}, S, t).$

Hence

$$\sum_i P(Q', q_i', S, t) \geq \sum_{i,n} P(Q, q_{in}, S, t).$$

But the right-hand side here is 1, and the left-hand side is ≥ 1. Hence each '$\geq$' in (i), for various i, must be an equality relation (or else the left-hand side is > 1). Therefore we have

(BSI)' $\quad P(Q', q_i', S, t) = \sum_n P(Q, q_{in}, S, t)$

$$= \sum_n |(f_{in}, f(S, t))|^2.$$

This result has several interesting corollaries. We include their proofs here for convenience, even though the machinery to prove them is developed in Chapter 4. Firstly we will prove that any linear combination of $\{f_{i1}, f_{i2}, \ldots\}$ must be an eigenvector of Q for value q_i (see Theorem 1, below), and vice versa (see Theorem 2, below). Thus we see that the eigenvectors of Q' fall into a set of mutually orthogonal closed linear subspaces $\{V_i\}$, one such subspace V_i for each possible value q_i'. And, as we would expect, the value of $P(Q', q_i', S, t)$ is independent of the choice of a basis set $\{f_{i1}, f_{i2}, \ldots\}$ from among the vectors in V_i. This is because if V_i is the set of all linear combination of $\{f_{i1}, f_{i2}, \dots\}$ as well as the set of all linear combinations of $\{f'_{i1}, f'_{i2}, \dots\}$ then it is easy to show (see Theorem 3 below) that

$$\sum_n |(f_{in}, f)|^2 = \sum_n |(f'_{in}, f)|^2.$$

Theorem 1. Let $f = \sum_n c_n f_{in}$, where $\sum |c_n|^2 = 1$.

Then $\sum_n |(f_{in}, f)|^2 = \sum_n |(f_{in}, \sum_m c_m f_{im})|^2$

Hence, since $(f_{in}, f_{im}) = 0$ for $n \neq m$,

$$\sum_n |(f_{in}, f_i)|^2 = \sum_n |(f_{in}, f_{in})|^2 |c_n|^2$$

$$= \sum_n |c_n|^2 = 1.$$

Hence, by (BSI)', we see that

$P(Q', q'_i, S, t) = 1$ if $f(S, t) = f$; and hence f is an eigenvector of Q'.

Theorem 2. Let f fall outside V_i, the set of linear combinations of the $\{f_{i1}, f_{i2}, \dots\}$, so that f is of the form $c_1 f_1 + c_2 f_2$, where f_1 is in V_i, f_2 is orthogonal to V_i, $c_2 \neq 0$ and $|c_1|^2 + |c_2|^2 = 1$. Clearly if $f(S, t) = f$ then $P(Q', q'_1, S, t)$ is $|c_1|^2$ which is < 1 (since $c_2 \neq 0$). Hence f is not an eigenvector of Q' for value q'_i.

Theorem 3. As for Theorem 2 let

(i) $f = \sum_n c_n f_{in} + f_2 = \sum_n c'_n f'_{in} + f_2$,

where $f_{in} = \sum_{n'} c'_{nn'} f'_{in'}$ (since V_i includes f_{in}).

But $(f_{in'}, f_{in''}) = \delta_{n'n''}$

Hence

(ii) $\sum_n \bar{c}'_{n'n} c'_{n''n} = \delta_{n'n''}$

Now we also have that

$$f'_{in} = \sum_{n'} c_{nn'} f_{in'}$$

And since $(f_{in'}, f'_{in}) = (f'_{in}, f_{in'})^*$ we have

(iii) $c_{nn'} = \bar{c}'_{n'n}$

Hence, by (ii) and (iii),

$$\sum_n c_{nn'} \bar{c}_{nn''} = \delta_{n'n''}$$

Hence, taking complex conjugates of both sides,

(iv) $\sum_n \bar{c}_{nn'} c_{nn''} = \delta_{n'n''}$

Now consider

$$\sum_n |(f'_{in}, f)|^2$$

$$= \sum_n \left(\sum_{n'} c_{nn'} f_{n'}, f \right) \left(f, \sum_{n''} c_{nn''} f_{n''} \right)$$

which, by (i),

$$= \sum_{n',n''} c_{n'} \bar{c}_{n''} \sum_n \bar{c}_{nn'} c_{nn''}$$

which, by (iv),

$$= \sum_{n',n''} c_{n'} \bar{c}_{n''} \delta_{n'n''}$$

$$= \sum |c_n|^2$$

$$= \sum_n |(f_{in}, f)|^2 \quad \text{QED.}$$

Note that these results can be used to retrospectively justify (A) (i.e. (i) at the beginning of this appendix), by showing how it implies various of the fundamental principles of QT such as (BSI)′.

Continuous-valued Physical Quantities

The key to understanding von Neumann's treatment of continuous-valued physical quantities is to realize what his overall strategy is from the start. In the case of non-degenerate discrete-valued physical quantities we associate a vector with each possible value; for degenerate discrete-valued physical quantities there is a subspace associated with each possible value. The various vectors associated with distinct possible values are orthogonal to one another. And (as we indicated above) the Hilbert space in which we locate these vectors is separable, i.e. it contains at most a countable infinity of mutually orthogonal vectors.

Clearly then something has to be changed in this formalism when we graduate to continuous-valued quantities (for which there are more than countably many different possible values). How are we to make the change? The problem lies not so much in thinking of some way to make the change, but rather in how to narrow down the range of options. One option is the Dirac–Jordan approach using delta-functions as approximate eigenfunctions, but which von Neumann rejects because of its lack of rigour (see (1955) pp. 24 ff).[1] Before discussing von Neumann's approach in some detail we shall describe (in slightly simplified form) what his approach is, divorced from the mathematical framework which makes it so plausible as the uniquely appropriate answer.

Von Neumann suggests that with every physical quantity Q we associate a family of projection operators—one for each possible value q of Q, with '$\hat{E}(q)$' denoting the projection for the possible value q. (By contrast in the special case where Q is discrete-valued we associated a vector, or more generally a closed linear subspace, with each possible value.) These projections satisfy the following formal properties (i), (ii), (iii), (iv)a, (iv)b:

(i) $\hat{E}(q)$ tends to $\hat{I}$ as q tends to plus infinity.
 $\hat{E}(q)$ tends to $\hat{0}$ as q tends to minus infinity.

($\hat{I}$ is the identity operator, i.e. $\hat{I}g = g$ for any g; and $\hat{0}$ is the zero operator, i.e.

---

[1] References in this Appendix to (1955) are to von Neumann (1955).

$\hat{0}g = 0$ for any g. Note that if the set of possible values is bounded above and below by q_0 and q_1 respectively then (i) becomes simply: $\hat{E}(q_1) = \hat{I}$ and $\hat{E}(q_0) = \hat{0}$.)

(ii) If $q \leq q'$ then $\hat{E}(q')\hat{E}(q) = \hat{E}(q)\hat{E}(q') = \hat{E}(q)$ (which is written as: $\hat{E}(q) \leq \hat{E}(q')$).

(iii) $\hat{E}(q)$ tends to $\hat{E}(q')$ as q tends to q' from above (i.e. $\hat{E}(q)f$ tends to $\hat{E}(q')f$ as q tends to q' from above, for any f), i.e. $\hat{E}$ is continuous from above.

Any set of projections $\{\hat{E}(q)\}$ which has the properties (i) to (iii) is said to be a resolution of the identity. Moreover the $\{\hat{E}(q)\}$ are supposed to act as bridges with observable reality by satisfying:

(iv)a $(\hat{E}(q'') - \hat{E}(q'))f(S, t) = f(S, t)$ iff Q has a value within the interval $(q', q'']$ for S at t with certainty, i.e. $P(Q, (q', q''], S, t) = 1$.

(Here '$(q', q'']$' denotes the semi-closed interval of values q for which $q' < q \leq q''$. Thus it is assumed $q'' > q'$.)

(iv)b Let $\hat{E}^0(q)$ be the limit of $\hat{E}(q')$ as q' tends to q from below. Then $(\hat{E}(q) - \hat{E}^0(q))f(S, t) \doteq f(S, t)$ iff Q has the value q for S at t with certainty, i.e. $P(Q, q, S, t) = 1$.

(Note (iv)b can be regarded as a kind of limiting case of (iv)a, in the limit as we allow $(q' q'']$ to become smaller by increasing q' up to q''. Note also that the semi-continuity of $\hat{E}$—as stated in (iii)—rather than full continuity—as stated in (iii)—is required by (iv)b if a quantity is ever to have a particular value with certainty.)

We must now show that the above suggestion does indeed represent a generalization of the rules of QT as so far construed in this book. In particular we must show that the rules (i)–(iv) accommodate the forms of representation for q-quantities we have already discussed, and in particular make room for discrete-valued quantities. Moreover we must provide a generalization of the Born interpretation which fits both the continuous and discrete-valued cases we have already discussed. The relevant generalized Born interpretation is:

(BSI)'' $P(Q, (q' q''], S, t) = ((\hat{E}(q'') - \hat{E}(q'))f(S, t), (\hat{E}(q'') - \hat{E}(q'))f(S, t))$
and $P(Q, q, S, t) = ((\hat{E}(q) - \hat{E}^0(q))f(S, t), (\hat{E}(q) - \hat{E}(q^0))f(S, t))$.

(Clearly this has (iv)a and (iv)b as special cases, and is indeed suggested by these.) We shall now show that (BSI)'' does indeed have the earlier forms of (BSI) as special cases.

From (BSI)'' we see the following. Consider a special case of a Q for which the $\{\hat{E}(q)\}$ are continuous at all q except a countable sequence of values $q_1, q_2, \ldots$ (although they are of course continuous from above at the various q_i—in accord with (iii)). And suppose that the $\{\hat{E}(q)\}$ are constant for all q between a given successive pair of members of the $\{q_i\}$, i.e. $\hat{E}(q) = \hat{E}(q_i)$ for any

q such that $q_i \leq q < q_{i+1}$. In that case it is clear that

(a) $P(Q, q, S, t) = 0$ for any $q \neq$ one of the q_i.

(b) $P(Q, q_i, S, t) = ((\hat{E}(q_i) - \hat{E}^0(q_i))f(S, t), (\hat{E}(q_i) - \hat{E}^0(q_i))f(S, t))$.

But $(\hat{E}(q_i) - \hat{E}^0(q_i))$ is still a projection operator (for proof see footnote[2]) and hence there is a closed linear subspace V_i for which $(\hat{E}(q_i) - \hat{E}^0(q_i)) = \hat{P}(V_i)$. Therefore (b) says:

$$P(Q, q_i, S, t) = (\hat{P}(V_i)f(S, t), \ \hat{P}(V_i)f(S, t)).$$

But $\hat{P}(V_i)f(S, t) = \Sigma_n (f_{in}, f(S, t))f_{in}$ for any set $\{f_{i1}, f_{i2}, \ldots\}$ which is a basis set for V_i and hence we trivially arrive at:

$$P(Q, q_i, S, t)$$
$$= \left(\sum_n (f_{in}, f(S, t))f_{in}, \sum_m (f_{im}, f(S, t))f_{im} \right)$$
$$= \sum_n \sum_m (f_{in}, f(S, t))^* (f_{im}, f(S, t)) (f_{in}, f_{im}),$$

which, since *ex hypothesi* $(f_{in}, f_{im}) = 0$ for $n \neq m$, and $= 1$ for $n = m$, reduces to:

$$\sum_n |(f_{in}, f(S, t))|^2,$$

which is just the (BSI)′ of Appendix 3.

Moreover from additional assumptions we shall make later the various V_i are seen to be orthogonal, i.e. each vector in V_i is orthogonal to each vector in V_j. Thus in this special case we have a physical quantity Q for which there is zero probability of observing it to have any value other than a member of some countably infinite set $\{q_i\}$, and there is a closed linear subspace V_i associated with each q_i in such a way as to satisfy (BSI)′ (the various V_i being orthogonal). Here then we have our discrete-valued physical quantities. They are discrete-valued in the (weak) sense that for any one of them there is only a countable set of possible values for which any one of the set can be observed (with non-zero probability). Moreover the old Born interpretation is satisfied by these discrete-valued quantities.

Thus our generalized Born interpretation satisfies the conditions we required above, viz. it provides surrogates for the old discrete-valued quantities (albeit somewhat weakened in form), and it reduces to the old Born interpretation for those discrete-valued quantities. It still remains, however,

---

[2] Consider $\hat{E}(q) - \hat{E}(q')$ for $q' < q$, and hence (by (ii)) $\hat{E}(q)\hat{E}(q') = \hat{E}(q')\hat{E}(q) = \hat{E}(q')$. Now we know that $\hat{E}^2 = \hat{E}$ is necessary and sufficient for $\hat{E}$ to be a projection. Hence

$$(\hat{E}(q) - \hat{E}(q'))^2 = \hat{E}(q)^2 - \hat{E}(q)E(q') - \hat{E}(q')\hat{E}(q) + \hat{E}(q')^2$$
$$= \hat{E}(q) + \hat{E}(q') - \hat{E}(q') - \hat{E}(q') = \hat{E}(q) - \hat{E}(q').$$

Hence the limit of $(\hat{E}(q) - \hat{E}(q'))$ as q' tends to q from below is a projection too.

to show that (BSI)″ gives the appropriate expression for $P(Q, (q', q''], S, t)$ for the already known cases where Q is continuous-valued (and hence that it succeeds in blending the continuous and discrete within the one formalism). This is in fact easily done (see (1955) pp. 198–9) as follows:

Proof   The resolution of identity $\{\hat{E}(x)\}$ which is associated with a position coordinate X with possible values $\{x\}$ (which for simplicity, we take as the position coordinate of some one-dimensional q-system) is taken to be the following:

$$\hat{E}(x)f(x') = f(x') \text{ for all } x' \leq x$$
$$= 0 \text{ for all } x' > x,$$

where the vector f is here taken as a function of position and hence of the possible values of X. (The rationale for this choice of $\hat{E}(x)$ is given later.) Hence

$$P(X, (x', x''], S, t) = ((\hat{E}(x'') - \hat{E}(x'))f(S, t), (\hat{E}(x'') - \hat{E}(x'))f(S, t))$$

Now let $f(x)$ be the value of $f(S, t)$ for argument x, and define $\bar{f}(x)$ as the value of the function $(\hat{E}(x'') - \hat{E}(x'))f(S, t)$ for argument x, i.e.

$$\bar{f}(x) = 0 \text{ for } x > x''$$
$$= f(x) \text{ for } x' \leq x \leq x''$$
$$= 0 \text{ for } x < x'$$

Then $P(X, (x', x''], S, t) = \int |\bar{f}(x)|^2 \, dx$
$$= \int |f(x)|^2 \, dx \text{ for } x \text{ in the interval } (x', x'']$$

and clearly that is exactly the result we want, since in differential form this says:

$$P(X, (x, (x + dx], S, t) = |f(x)|^2 \, dx$$

which is just the old form of the Born interpretation for a continuous position coordinate. QED

Now let me try and relate the above treatment of continuous valued quantities more closely to von Neumann's. My treatment differs from von Neumann's in that I have avoided representing physical quantities by operators until the last possible moment. Von Neumann takes it as central that any physical quantity Q (for system S) is represented by a closed linear hermitean operator $\hat{Q}$ (on the Hilbert space $\mathcal{H}$ for S), and that this feature is preserved even when we consider continuous-valued quantities. In other words we do not need to move out of $\mathcal{H}$ (e.g. by 'rigging it', as Dirac and Jordan would have us do) when we consider continuous-valued quantities. (By continuous-valued quantities I mean any physical quantities which have an uncountable number of possible values even though some of their possible values may be discrete, i.e. not part of a continuum.) Moreover he takes it as central that it is the eigenvalue problem which tells us both what possible values there are for Q and ultimately, via the Born interpretation, what their relevant probabilities are. In the discrete case the eigenvalue problem is simply stated. It asks:

(Q) What basis set $\{f_i\}$ of $\mathcal{H}$ and what set of numbers $\{q_i\}$ have the property that $\hat{Q}f_i = q_i f_i$ for every i?

The various $\{q_i\}$ are then taken as the possible values of Q. (They are the eigenvalues of $\hat{Q}$.) And the Born interpretation gives us the relevant probabilities via (BSI)'. Thus by solving the purely mathematical problem (Q), i.e. by solving the eigenvalue problem for $\hat{Q}$, we get the physical information we want. Moreover there is a guarantee that if $\hat{Q}$ is closed, linear and hermitean and $\mathcal{H}$ is finite dimensional then a solution to the eigenvalue problem is forthcoming. That is it is provable that for any linear hermitean operator on a finite dimensional Hilbert space there is a complete orthonormal (basis) set of eigenvectors, or at least this is so if the operator has been extended to cover all of the space in such a way that its closed linear hermitean nature is preserved. Such an extension is always possible in a finite dimensional space. In what follows I shall assume that we always deal with such extended operators (see (1955) pp. 107 and 167–9).

But what about in an infinite (countably infinite) dimensional Hilbert space? In that case it turns out that there is no answer to (Q) in certain cases, i.e. there may be no complete orthonormal vectors $\{f_i\}$ for which $\hat{Q}f_i = q_i f_i$ for various q_i. At best there may be an incomplete orthonormal set $\{f_i\}$ with this property, and sometimes not even that. This is in part because the extension of an operator to cover the whole space, which is possible with finite-dimensional operators, may not be possible in infinite-dimensional spaces where the maximal operators, which have been maximally extended consistent with preserving their closed linear hermitean natures, may still not cover the whole space (see (1955) pp. 153, 154, 167–70). Moreover these are not just pathological cases, but include many of the cases which are of physical significance.

The problem which von Neumann then faces is this: how are we best to generalize this question (Q) to a weaker question which can be answered even if $\mathcal{H}$ is infinite-dimensional? He solves this problem by reformulating the question (Q) as a question which is logically equivalent to (Q) in the case that $\mathcal{H}$ is finite dimensional, but which is easily (and obviously) generalizable to the case where $\mathcal{H}$ is infinite dimensional. We shall now discuss this reformulation which hinges upon the following basic theorem for Hilbert spaces:

If $\{f_i\}$ is a complete orthonormal set of eigenvectors for $\hat{Q}$, so that $\hat{Q}f_i = q_i f_i$ for all i then $\hat{Q} = \Sigma q_i \hat{P}(f_i)$,

which was proved in Appendix 1.

Let $\{f_i\}$ be a complete orthonormal set of eigenvectors for a closed linear hermitean $\hat{Q}$ in a finite-dimensional Hilbert space. Introduce a continuous index 't' which varies between minus infinity and plus infinity and define

$$\hat{E}(t) = \sum_{i: q_i \leq t} \hat{P}(f_i),$$

which is the sum of the $\hat{P}(f_i)$ over all those i for which $q_i \leq t$. And pick out a countable set of values $\{t_1, t_2, \ldots, t_m, \ldots\}$ which includes all the $\{q_i\}$ but which has a separation between successive t_m which is $\leq d$ (thus d will have to be $\leq$ the minimum separation of the $\{q_i\}$). Clearly $\hat{E}(t_m) - \hat{E}(t_{m-1}) = 0$ if $t_m \neq$ one of the $\{q_i\}$, and $\hat{E}(t_m) - \hat{E}(t_{m-1}) = \hat{P}(f_i)$ if $t_m = q_i$ (since there can be no other of the $\{q_i\}$ between t_m and t_{m-1}). Thus $\hat{Q} = \Sigma_m (\hat{E}(t_m) - \hat{E}(t_{m-1})) t_m$.

By letting the spacing d become vanishingly small the right-hand sum becomes a Lebesgue–Stieltjes integral:

$$= \int d\hat{E}(t)\, t.$$

Moreover we easily see that $\{\hat{E}(t)\}$ satisfies the following conditions, which are a slight strengthening of the conditions for being a resolution of the identity (defined above):

> For t large enough $\hat{E}(t)$ is just $\Sigma \hat{P}(f_i)$, which (given that the set $\{f_i\}$ is complete) is just $\hat{I}$,[3] and for t small enough $\hat{E}(t)$ is $\hat{O}$. Also it is clear that $(\hat{E}(t)\hat{E}(t')) = (\hat{E}(t')\hat{E}(t)) = \hat{E}(t)$ if $t \leq t'$, and the $\{\hat{E}(t)\}$ are continuous from above at any t.

Any family $\{\hat{E}(t)\}$ which satisfies these conditions we call 'a strong resolution of the identity'.

Moreover it is provable that for any $\{\hat{E}(t)\}$ which is a strong resolution of the identity and for which $\hat{Q} = \int d\hat{E}(t)t$, there is a discontinuity in the $\{\hat{E}(t)\}$— say at $t = q_i$—iff q_i is an eigenvalue of $\hat{Q}$ and each vector f in the subspace, onto which $(\hat{E}(q_i) - \hat{E}^0(q_i))$ is a projector, is an eigenvector of $\hat{Q}$ for eigenvalue q_i, i.e. $\hat{Q}f = q_i f$ (see (1955) p. 123). Thus we have a restatement of the above eigenvalue problem (Q) for $\hat{Q}$:

(Q′) What strong resolution of the identity $\{\hat{E}(t)\}$ is there for which $\hat{Q} = \int d\hat{E}(t)t$?

Once we determine such an $\{\hat{E}(t)\}$ we can then solve the original eigenvalue problem (Q) (by considering the discontinuities of $\{\hat{E}(t)\}$, and vice versa.) There is of course a question of the existence of the $\{\hat{E}(t)\}$. For finite dimensional spaces this last question is trivial, because the $\{\hat{E}(t)\}$ always exist if $\hat{Q}$ is defined everywhere—see below—and, as we have already indicated, this can always be arranged by extending $\hat{Q}$ in such a way that its closed hermitean linear nature is preserved. In the case of infinite dimensional spaces

---

[3] I.e. $(\Sigma \hat{P}(f_i))g = \Sigma(\hat{P}(f_i)g) = \Sigma(f_i, g)g = g$.

however this is not always possible, i.e. in the case where our Hilbert space $\mathcal{H}$ is infinite-dimensional we find that at least for some $\hat{Q}$ there is no resolution $\{\hat{E}(t)\}$ (and *a fortiori* no strong resolution of the identity) for which $\hat{Q} = \int d\hat{E}(t) t$. In part this is because some of the $\hat{Q}$ are not defined everywhere in $\mathcal{H}$ even after they have been extended maximally. Nevertheless one might expect there to be a resolution of the identity $\{\hat{E}(t)\}$ for which we can write $\hat{Q}f = (\int d\hat{E}(t)t)f$ for at least those f in $\mathcal{H}$ for which $\hat{Q}f$ exists in $\mathcal{H}$. Moreover one would like it to be the case that the only reason for the non-existence of $\hat{Q}f$ in $\mathcal{H}$ is that $\hat{Q}$ is unbounded in its action on f, i.e. $|\hat{Q}f|$ is infinite; in other words that there are no arbitrary restrictions on the domain of $\hat{Q}$—it has been maximally extended. Or, as von Neumann puts it (1955, pp. 118–19), we expect there to be an $\{\hat{E}(t)\}$ for which $(g, \hat{Q}f) = \int td(g, \hat{E}(t)f)$ for any g, f in $\mathcal{H}$ for which $\int t^2 d(|\hat{E}(t)f|^2)$ is finite and $\int t^2 d(|\hat{E}(t)f|^2)$ is infinite for any other f.[4]

In fact this expectation is realized, except for a pathological class of operators which are of a very restricted type and generally of no physical interest (see (1955) p. 168). That is any closed hermitean linear and maximal operator $\hat{Q}$ on $\mathcal{H}$ which is of physical significance is also hypermaximal in that there is a resolution of the identity $\{\hat{E}(t)\}$ for which $\hat{Q}f = (\int d\hat{E}(t)t)f$ for any f for which $\hat{Q}f$ is in $\mathcal{H}$.

The whole point now is this. If $\mathcal{H}$ is infinite dimensional then the $\{\hat{E}(t)\}$ may be continuous at all t, or, more generally, the discontinuities, if any, may be such that the set of non-zero projections $\{\hat{E}(t) - \hat{E}^0(t)\}$ at the discontinuities is not complete. In a finite-dimensional space of course this cannot happen and the questions (Q) and (Q)' are then equivalent. But in an infinite-dimensional space this equivalence breaks down. The $\{\hat{E}(t)\}$ may be continuous everywhere and hence there may be no eigenvalues or eigenvectors at all, i.e. no q, f for which $\hat{Q}f = qf$.[5] Von Neumann then distinguishes three different sorts of values for t. There are those values t for which the $\{\hat{E}(t)\}$ are discontinuous at t. These are the eigenvalues of $\hat{Q}$, and make up the point spectrum. Then there are the values t for which although the $\{\hat{E}(t)\}$ are continuous at t they are not constant in the neighbourhood of t (i.e. $d\hat{E}(t) \neq 0$). These make up the continuous spectrum.[6] Then there are the other t-values at which the $\{\hat{E}(t)\}$ are continuous and constant. It is the continuous spectrum plus point spectrum which von Neumann then takes to make up the possible

---

[4] Von Neumann's formulation of this condition can be seen to be essentially the same as mine once one realizes that the condition that $\hat{Q}f$ is in $\mathcal{H}$ is just the condition that $(\hat{Q}f, \hat{Q}f)$ (which is just $|\hat{Q}f|^2$) is finite; and if $\hat{Q} = \int d\hat{E}(t)t$ this last condition gives precisely the von Neumann condition that $\int d|\hat{E}(t)f|^2 t^2$ is finite.

[5] As we indicated above, the discontinuities of the $\{\hat{E}(t)\}$ are in 1:1 correspondence with the distinct eigenvalues and their eigenspaces.

[6] It is easily proved that if the $\{\hat{E}(t)\}$ are continuous over the interval $(t', t'']$, then $\hat{E}(t) - \hat{E}(t'')$ is a projection onto a subspace orthogonal to any $\hat{E}(t)$ for which there is a discontinuity in the $\{\hat{E}(t)\}$ at t. Thus the existence of a continuous spectrum implies the incompleteness of the set of eigenvectors (see (1955) pp. 123–4).

values for the physical quantity Q which is represented by $\hat{Q}$. This fits in with the (BSI)″ which I gave earlier on, i.e. for any point q in the continuous spectrum, $P(Q,q,S,t)=0$ but nevertheless $P(Q,(q', q''], S, t)$ may be non-zero for any $(q', q'']$ containing q however small the interval is. For points in neither spectrum however, not even this last property holds, and in this strong sense they are not possible values. Indeed their only role seems to be in enabling the construction of $\{\hat{E}(t)\}$. Although formally they are not even necessary for that, i.e. we can eliminate the values of t which are in neither spectrum and still have a resolution of the identity for which $\hat{Q}f=\int d\hat{E}(t)\,t$, i.e. if the $\{\hat{E}(t)\}$ are constant in some neighbourhood of value t then those values do not contribute to the integral. In other words with no loss of generality we can insist that the $\{\hat{E}(t)\}$ are non-constant in the neighbourhoods of all values of t, although the set of such values may then be gappy, i.e. not form a continuum.

In sum then we have seen how von Neumann extracts information about the possible values of a physical quantity Q, even for continuous-valued quantities, from the generalized eigenvalue problem for the representative operator $\hat{Q}$. And the relevant statistical information about Q's measured values then follows with the help of (BSI)″.

The question now is this. Why bother with the long and complex detour via $\hat{Q}$? Why not associate a resolution of identity directly with Q, as we did initially? From a logical point of view it turns out that there is no reason for not doing this. The simple assumption that there is a resolution of identity associated with each physical quantity Q which satisfies the conditions (iv)a and (iv)b given above produces a version QT which is equivalent to von Neumann's conceptually more complex approach which associates an hypermaximal (linear closed hermitean) operator with each Q and then solves the eigenvalue problem for $\hat{Q}$ in order to ascertain the possible values and the states which make them certain.

The equivalence arises because it is provable that for each resolution of identity $\{\hat{E}(t)\}$ there is a linear closed hermitean operator $\hat{Q}$ for which von Neumann's condition (see above):

$(g, \hat{Q}f)=\int t d(g, \hat{E}(t)f)$ for any g,f in $\mathscr{H}$ for which $\int t^2 d(|\hat{E}(t)f|^2)$ is finite, and
$\int t^2 d(|\hat{E}(t)f|^2)$ is infinite for any other f in $\mathscr{H}$

is satisfied. (For proof, see reference in (1955), footnote 103, and in particular von Neumann (1929).) This proof is a generalization to the continuous case of the theorem proved above that for every basis set $\{f_i\}$ and set of possible values $\{q_i\}$ there is a $\hat{Q}$ such that $\hat{Q}f_i=q_i f_i$.

There are nevertheless significant advantages to the operator representation. In particular there is the advantage already discussed in Appendix 2, viz. that the operator representatives of physical quantities correspond to their classical analogues via the heuristically significant principle (Corresp.). And we now see the further advantage that it is the operator representation of physical quantities which is preserved in making the generalization from discrete-valued to continuous-valued physical quantities, i.e. it is via a generalization of the eigenvalue problem to the infinite-dimensional case that we can see how to generalize the treatment of physical quantities within QT to include quantities which are continuous-valued.

The Density Operator Formalism

The density operator formalism is convenient when we generalize the notion of a state in QT to allowing that for each S and t there is a set of mutually orthogonal vectors f_x for various $x = 1, 2, \ldots$, and there are associated probabilities p_x where $\Sigma p_x = 1$, for which for any Q, i,

(BSI)* $\quad P(Q, i, S, t) = \sum_x p_x \sum_n |(f_{in}, f_x)|^2.$

In this case the state of S at t is said to be 'mixed', and is characterized by the N-tuple of pairs $\{p_x, f_x\}$.

In order to put (BSI)* into the language of the density operator we need to introduce the trace operation. The trace operation is a function which associates either a unique real number or infinity with every operator on $\mathscr{H}$. It is defined q.v.:

$$\mathrm{Tr}\, \hat{Q} = \sum (f_i, \hat{Q} f_i) \text{ for any basis set } \{f_i\} \text{ in } \mathscr{H}.$$

Of course for this to be a viable definition it must be that $\Sigma(f_i, \hat{Q} f_i)$ takes on a value which is independent of the choice of the basis set $\{f_i\}$. This is easily shown to be the case:

Proof Let $\{f'_i\}$ and $\{f_i\}$ both be basis sets for $\mathscr{H}$.
Then

$$f'_i = \sum_{i'} c_{ii'} f_{i'} \text{ for } c_{ii'} = (f_{i'}, f'_i)$$

and, of course,

$$f_i = \sum_{i'} d_{ii'} f'_{i'} \text{ for } d_{ii'} = (f'_{i'}, f_i),$$

Clearly since $(f_{i'}, f'_i) = (f'_i, f_{i'})^*$ we have $c_{ii'} = d^*_{i'i}$.

Then $\sum_i (f'_i, \hat{Q}f'_i) = \sum_i \sum_{i''} c_{ii''}(f'_i, \hat{Q}f_{i''})$

$$= \sum_{i'} \sum_{i''} (f_{i'}, \hat{Q}f_{i''}) \sum_i c^*_{ii'} c_{ii''}$$

$$= \sum_{i'} \sum_{i''} (f_{i'}, \hat{Q}f_{i''}) \sum_i d_{i'i} d^*_{i''i}$$

$$= \sum_{i'} \sum_{i''} (f_{i'}, \hat{Q}f_{i''})(f_{i'}, f_{i''})^*$$

$$= \sum_i (f_i, \hat{Q}f_i), \text{ since } (f_{i'}, f_{i''}) = 0 \text{ for } i' \neq i'' \text{ and } = 1 \text{ for } i' = i''.$$

QED.

(The sums in this last proof were treated as if they were finite. This treatment is easily justified in the case of infinite sums by using the closure of $\hat{Q}$.)

Moreover we easily see that $\mathrm{Tr}(c_1\hat{Q}_1 + c_2\hat{Q}_2) = c_1\mathrm{Tr}\,\hat{Q}_1 + c_2\mathrm{Tr}\,\hat{Q}_2$; and using the closure property for operators we see that $\mathrm{Tr}\,\Sigma\,c_i\,\hat{Q}_i = \Sigma\,c_i\,\mathrm{Tr}\,\hat{Q}_i$. Also trivially we see that $\mathrm{Tr}\,\hat{P}(f)\,\hat{P}(g) = |(f,g)|^2$.

Proof Let $\{g_1, g_2, g_3, \dots\}$ be some basis set in $\mathscr{H}$, with $g_1 = g$. Then

$$\mathrm{Tr}\,\hat{P}(f)\hat{P}(g) = \sum_i (g_i, \hat{P}(f)\hat{P}(g)g_i)$$

$$= \sum_i (g_i, \hat{P}(f)(g, g_i)g)$$

$$= (g, \hat{P}(f)g) \text{ (since } g = g_1)$$

$$= (g, (f, g)f)$$

$$= (f, g)(g, f) = |(f, g)|^2. \text{ QED}$$

Now we define the density operator $\hat{W}(S, t)$ for S at t as follows:

Defn $\hat{W}(S, t) = \Sigma\,p_x\hat{P}(f_x)$ if the state of S at t is $\{p_x, f_x\}$.

Thus the state of S at t is pure iff $\hat{W}(S, t)$ is a projection operator $\hat{P}(f)$ for some f.

Then we can prove:

(BSI)** $P(Q, i, S, t) = \mathrm{Tr}\,\hat{W}(S,t)\hat{P}(Q,i)$ (where $\hat{P}(Q, i)$ as above is the projection operator onto the eigenspace of Q for the ith distinct value q_i).

Proof Let S at t be in the mixed state $\{p_x, f_x\}$. Then

$$\mathrm{Tr}\,\hat{W}(S,t)\,\hat{P}(Q,i) = \mathrm{Tr}\,\sum_x p_x\hat{P}(f_x)\sum_n \hat{P}(f_{in}) \text{ (by Defn.)}$$

$$= \sum_x p_x \sum_n \mathrm{Tr}\,\hat{P}(f_x)\hat{P}(f_{in})$$

$$= \sum_x p_x \sum_n |(f_{in}, f_x)|^2$$

$$= P(Q, i, S, t) \text{ (by (BSI)*). QED}$$

Note that if $\langle Q \rangle$, the expectation value of Q for S at t, is defined as $\Sigma\, q_i P(Q, i, S, t)$ then from (BSI)** and the fact derived above that $\hat{Q} = \Sigma\, q_i \hat{P}(Q, i)$ it follows that $\langle Q \rangle = \mathrm{Tr}\ \hat{W}(S, t)\hat{Q}$. This is yet another variant of the Born interpretation which is to be commonly found in the literature.

As in Chapter 4 it follows from (Princ) and (Princ)′ and (A) that

(Bohr)′ If $\hat{W}(S, t) = \Sigma p_n \hat{P}(f_n)$ where f_n is an eigenvector of non-degenerate Q for value q_n and $q_n \neq q_m$ for all $n \neq m$ then there is probability p_n that Q has value q_n (determinately) in S at t for all n.

But we can now also prove that

(Bohr)″ If $\hat{W}(S, t) = \Sigma p_i \hat{P}(f_i)$ where $\{f_i\}$ is a complete orthonormal set of eigenvectors of the discrete-valued but possibly degenerate Q then the probability that Q has value q in S at t is $(p_n + p_{n+1} + \ldots + p_m)$ if $f_n, f_{n+1}, \ldots, f_m$ are those members of $\{f_i\}$ which are eigenvectors of Q for possible value q.

The proof makes use of a further principle (an extension of (A)):

(A)′ If Q has eigenvectors $f_1, f_2, \ldots$ for eigenvalue q and the non-degenerate Q' has eigenvector f_1 for eigenvalue q'_1, and has eigenvector f_2 for eigenvalue $q'_2, \ldots$ then $P(\mathrm{val}\, Q, q, S, t) \geqslant P(\mathrm{val}\, Q', q'_1, S, t) + P(\mathrm{val}\, Q', q'_2, S, t) + \ldots$.

(This can be seen as an implication of the modal principle that Q has value q whenever Q' has one of the values $q_1, q_2, \ldots$)

Proof (of (Bohr)″
   Let $\hat{W}(S, t) = \Sigma\, p_i \hat{P}(f_i)$ where each f_i is an eigenvector of Q but where Q may be degenerate. We arrange the $\{f_i\}$ so that $f_n, f_{n+1}, \ldots f_m$ are the eigenvectors of Q for value q. We assume (as part of assuming the existence of 'enough' physical quantities) that there is a non-degenerate Q' for which f_i is an eigenvector of Q' for value q'_i and hence (by (Princ)″) there is probability p_i that Q' has value q'_i in S at t. But $P(\mathrm{val}\, Q, q, S, t) \geq P(\mathrm{val}\, Q', q_n', S, t) + P(\mathrm{val}\, Q', q_{n+1}, S, t) \ldots + P(\mathrm{val}\, Q', q_m, S, t)$ (by (A)′). And the right-hand side probability is $p_n + p_{n+1} + \ldots + p_m$. Hence the probability of Q having value q in S at $t \geqslant p_n + \ldots + p_m$ too. Since $\Sigma\, p_i = 1$ and $\Sigma_q P(\mathrm{val}\, Q, q, S, t) = 1$ too, it follows that the last inequality must be an equality. QED.

APPENDIX 6

The Density Operator of Sub-systems

We shall here introduce a theorem which details the relation between the density operator of a system and the density operator of a subsystem. If S_1, S_2 are two q-systems then their combination (or 'join') is denoted by '$S_1 + S_2$'. (This notion of a combination of two systems functions as a primitive notion in QT.) The combination of two q-systems is itself taken to be a q-system, with its own Hilbert space. In particular $\mathscr{H}(S_1 + S_2)$ is taken to be the 'product-space' $\mathscr{H}(S_1) \times \mathscr{H}(S_2)$. We shall now explain what this 'product space' is.

Consider an N_1-dimensional Hilbert space $\mathscr{H}_1$. Then any vector f^1 in $\mathscr{H}_1$ can be represented by a sequence of N_1 numbers $(c_1, c_2, \dots)$—where c_i is (f_i^1, f^1), $\{f_i^1\}$ being some fixed basis set in $\mathscr{H}_1$. Similarly any vector f^2 in $\mathscr{H}_2$, an N_2-dimensional Hilbert space, can be represented by N_2 numbers $(d_1, d_2, \dots)$, where $d_j = (f_j^2, f^2)$, $\{f_j^2\}$ being some fixed basis set in $\mathscr{H}_2$. We then construct an $N_1 \times N_2$ dimensional Hilbert space $\mathscr{H}$ in which each vector f in $\mathscr{H}$ can be represented by a sequence of $N_1 \times N_2$ numbers $(c_1, c_2, \dots)$, where $c_i = (f_i, f)$ and $\{f_i\}$ is a fixed basis set in $\mathscr{H}$. For any f^1 in $\mathscr{H}_1$ and f^2 in $\mathscr{H}_2$ there is a vector in $\mathscr{H}$, which we denote $f^1 \times f^2$, such that $f^1 \times f^2$ is represented by the sequence of numbers $(c_1 \times d_1, c_1 \times d_2, \dots, c_1 \times d_{N_1}, c_2 \times d_1, \dots)$ if f^1 is represented by $(c_1, c_2, \dots)$ and f^2 is represented by $(d_1, d_2, \dots)$. Then the closure of the set of all such vectors is defined as $\mathscr{H}_1 \times \mathscr{H}_2$—the product of $\mathscr{H}_1$ with $\mathscr{H}_2$.[1] (This notion of a product space is easily generalized to the case where $\mathscr{H}_1$ or $\mathscr{H}_2$ are infinite-dimensional.)

For our purposes the following two easily proved theorems are important: $(f^1 \times f^2, g^1 \times g^2) = (f^1, g^1)(f^2, g^2)$; and $(f^1 + g^1) \times f^2 = (f^1 \times f^2) + (g^1 \times f^2)$. And note also the following law of QT:

(L_1) If $S_1 + S_2$ at t is in a pure state $f_1 \times f_2$ then so are S_1 and S_2, and $f(S_1, t) = f_1$ and $f(S_2, t) = f_2$.

(Note that this presupposes that $\mathscr{H}(S_1 + S_2) = \mathscr{H}(S_1) \times \mathscr{H}(S_2)$.)

And now finally we can prove the central theorem:

Theorem   If $f(S_1 + S_2, t) = \Sigma c_i f_i^1 \times f_i^2$ where $(f_i^1, f_{i'}^1) = (f_i^2, f_{i'}^2) = 0$ for any $i \neq i'$ (and $= 1$ for $i = i'$), then $\hat{W}(S_1, t) = \Sigma |c_i|^2 \hat{P}(f_i^1)$.

---

[1] I.e. $\mathscr{H}$ is got by taking all linear combinations of all such vectors.

Before proving this theorem it should be pointed out that it is actually a little more general in application than it looks. The assumption that $f(S_1 + S_2, t)$ $= \Sigma c_i f_i^1 \times f_i^2$ for some mutually othogonal $\{f_i^1\}$, and some mutually orthogonal $\{f_i^2\}$, is actually quite general. (See Jauch (1968) p. 179 for a discussion of this. It is a trivial consequence of the theorem that any bounded hermitean operator can be diagonalized.)

The proof of the theorem is straightforward once one realizes that any physical quantity Q_1 in S_1 will also be a physical quantity in any more comprehensive system $S_1 + S_2$. (This last assumption is discussed in the final section of Chapter 9). Thus the momentum of a particle S will also be a physical quantity for any system which is a combination of particles which includes S. Because of this, one and the same quantity Q may be represented by many different sets of eigenvectors (and hence operators), depending on which system it is being referred to and hence depending on the Hilbert space in which the eigenvectors are to be located. There is the following simple relation between these sets of eigenvectors:

Lemma   Let f_i^1 be any eigenvector of Q in the Hilbert space $\mathscr{H}_1$ for possible value q_i. Then the eigenvectors for Q in $\mathscr{H}_1 \times \mathscr{H}_2$ for possible value q_i consist of the set of all vectors $\{f_i^1 \times f^2\}$ for any vector f^2 in $\mathscr{H}_2$ and any f_i^1 which is an eigenvector of Q in $\mathscr{H}_1$ for possible value q_i.

Proof   Let $f(S_1 + S_2, t) = f_i^1 \times f^2$. Then clearly, by the above (L_1), $f(S_1, t) = f_i^1$ and hence Q has possible value q_i with certainty; and so $f_i^1 \times f^2$ is an eigenvector of Q for possible value q_i. Moreover it is easy to see that the set of vectors of form $\{f_i^1 \times f^2\}$ for varying i includes a complete set of vectors in $\mathscr{H}_1 \times \mathscr{H}_2$ and hence that $V(Q, q_i)$ in $\mathscr{H}_1 \times \mathscr{H}_2$ can include no more than the vectors of form $f_i^1 \times f^2$ for any i. QED

Now we can prove the theorem:

Proof   Let $f(S_1 + S_2, t) = \Sigma c_i f_i^1 \times f_i^2$ where $(f_i^1, f_{i'}^1) = 0$ for $i \neq i'$, and where with no loss of generality we can assume $\{f_i^2\}$ to be a basis set in $\mathscr{H}_2$. Let Q_1 be some physical quantity for S_1, and hence for $S_1 + S_2$ too. Suppose $\{g_i^1\}$ is a basis set of eigenvectors for Q_1, where Q_1 is non-degenerate when considered as a quantity for S_1 (i.e. $q_i^1 \neq q_{i'}^1$ for $i \neq i'$). Now from *Lemma* it follows that a basis set for $V(Q_1, q_i)$ in $\mathscr{H}_1 \times \mathscr{H}_2$ is $\{g_i^1 \times f_{i'}^2\}$ for varying i'. Thus, by (BSI)' of Appendix 3,

$$P(Q_1, q_i, S_1 + S_2, t) = \sum_{i'} |(g_i^1 \times f_{i'}^2, \sum c_{i''} f_{i''}^1 \times f_{i''}^2)|^2$$

$$= \sum_{i'} |\sum_{i''} c_{i''} (g_i^1 \times f_{i'}^2, f_{i''}^1 \times f_{i''}^2)|^2$$

$$= \sum_{i'} |\sum_{i''} c_{i''} (g_i^1, f_{i''}^1)(f_{i'}^2, f_{i''}^2)|^2$$

But

$(f_{i'}^2, f_{i''}^2) = 0$ for $i' \neq i''$.

Hence

$$= \sum_{i'} |c_{i'}(g_i^1, f_{i'}^1)|^2$$

$$= \sum_{i'} |c_{i'}|^2 |(g_i^1, f_{i'}^1)|^2.$$

Since $P(Q_1, q_i, S_1 + S_2, t) = P(Q_1, q_i, S_1, t)$, we then see that

$\hat{W}(S_1, t) = \sum |c_{i'}|^2 \hat{P}(f_{i'}^1)$. QED

Note that this theorem constitutes the non-trivial part of a more general theorem that $\hat{W}(S_1, t) = \mathrm{Tr}_2 \, \hat{W}(S_1 + S_2, t)$, where Tr_2 is the trace operation in $\mathscr{H}(S_2)$ as defined in Appendix 5. This is the von Neumann rule for the density operator of sub-systems:

(Von Neumann) $\hat{W}(S_1, t) = \mathrm{Tr}_2 \, \hat{W}(S_1 + S_2, t)$.[2]

---

[2] See von Neumann (1955) p. 425. Also for a discussion of this see Jauch (1968) sect. 11.8.

The Product of Pure States

'If S_1 is in the pure state f and S_2 is in the pure state g then $S_1 + S_2$ is in the pure state $f \times g$' follows trivially from the principle (Von Neumann) (end of Appendix 6) via the following theorem:

Theorem If $\hat{W}$ is a density operator on $\mathcal{H}_1 \times \mathcal{H}_2$ and $\mathrm{Tr}_2 \hat{W} = \hat{P}(f)$ where f is in $\mathcal{H}_1$ and $\mathrm{Tr}_1 \hat{W} = \hat{P}(g)$ where g is in $\mathcal{H}_2$, then $\hat{W} = \hat{P}(f) \times \hat{P}(g)$.

Proof For some basis sets $\{f_i\}$ and $\{g_j\}$ in $\mathcal{H}_1$ and $\mathcal{H}_2$ respectively

$$\hat{W} = \sum_{i,j} p_{ij} \hat{P}(f_i \times g_j).$$

(This is because any basis set in $\mathcal{H}_1 \times \mathcal{H}_2$ is of the form $\{f_i \times g_j\}$). Moreover, since $\hat{W}$ is a density operator, $p_{ij} \geq 0$ and $\Sigma_{i,j} p_{ij} = 1$. Hence

$$\mathrm{Tr}_1 \hat{W} = \sum_{i,j} p_{ij} \hat{P}(g_j).$$

But *ex hypothesi*

$$\mathrm{Tr}_1 \hat{W} = \hat{P}(g).$$

Hence

$$\sum_{i,j} p_{ij} \hat{P}(g_j) = \hat{P}(g).$$

Now operating with both sides on the vector g gives

$$\sum_{i,j} p_{ij}(g_j, g)\, g_j = g,$$

where, since $\{g_j\}$ is a basis set,

$$g = \sum_{j} (g_j, g) g_j.$$

Hence

$$\sum_{j} (g_j, g) g_j \left(\left(\sum_{i} p_{ij} \right) - 1 \right) = 0.$$

Hence, for each j, either $(\Sigma_j p_{ij}) - 1 = 0$ or $(g_j, g) = 0$.
Hence either $\Sigma_i p_{ij} = 1$ for some value of j or $(g_j, g) = 0$ for all j.

Since the second disjunct is not possible, we see that $\Sigma_i\, p_{ij} = 1$ for some value of j, say j^*. Since $p_{ij} \geq 0$ and $\Sigma_{i,\, j}\, p_{ij} = 1$ it follows that $p_{ij} = 0$ for all $j \neq j^*$.

And similarly, by considering $\mathrm{Tr}_2\, \hat{W}$, we prove that, for some i^*, $p_{ij} = 0$ for all $i \neq i^*$.

Hence

$$\hat{W} = \hat{P}(f_{i^*} \times g_{j^*})$$
$$= \hat{P}(f_{i^*}) \times \hat{P}(g_{j^*}),$$

and we see that g must equal g_{j^*}. Similarly we see that $f = f_{i^*}$. QED

APPENDIX 8

Impossibility of Reduction

Let $\hat{U}$ be a linear unitary operator. The question is whether for each choice of $\{c_i\}$ (for which $\Sigma |c_i|^2 = 1$) there is a member of the set $\{f_i\}$, f_j say, for which

(i) $\hat{U} \Sigma c_i f_i = f_j.$

We will assume (i), and show it implies a contradiction. It is essential for the proof that (i) holds for an arbitrary choice of $\{c_i\}$, although what value j takes may depend on the $\{c_i\}$.

Operating on both sides of (i) with $\hat{U}^*$ gives

(ii) $\Sigma c_i f_i = \hat{U}^* f_j$ (since $\hat{U}^*$ is unitary);

and hence, for all i,

(iii) $c_i = (f_i, \hat{U}^* f_j).$

Now choose $\{c_i\}$ for which $c_1 = 1$ and hence (since $\Sigma |c_i|^2 = 1$) $c_i = 0$ for $i \neq 1$. Let the j-value for which the equation (i) holds for that particular choice of $\{c_i\}$ be $j(1)$. Then, by (iii),

$(f_i, \hat{U}^* f_{j(1)}) = 0$ for all $i \neq 1$, and

$(f_1, \hat{U}^* f_{j(1)}) = 1.$

Now let $j(n)$ be the j-value for which equation (i) holds for the choice of $\{c_i\}$ for which $c_n = 1$ (and hence $c_m = 0$ for all $m \neq n$). Then we similarly get

(iv) $(f_i, \hat{U}^* f_{j(n)}) = 0$ for all $i \neq n$, and

(iv)′ $(f_n, \hat{U}^* f_{j(n)}) = 1.$

Moreover it is clear that $j(n) \neq j(m)$ for $m \neq n$,[1] so that $\{f_{j(n)}\} = \{f_i\}$.

Now consider $\hat{U}(f_1 + f_2)/\sqrt{2} = (\hat{U} f_1 + \hat{U} f_2)/\sqrt{2}$ (since $\hat{U}$ is linear). *Ex hypothesi* this is equal to one of the $\{f_i\}$, and hence one of the $\{f_{j(n)}\}$ (since, as we have just shown, $\{f_{j(n)}\} = \{f_i\}$), let it be $f_{j(n)}$, say. Thus

(v) $(\hat{U} f_1 + \hat{U} f_2)/\sqrt{2} = f_{j(n)}.$

---

[1] Suppose to the contrary that $j(n) = j(m)$ for $m \neq n$. Then, from (iv), $(f_m, \hat{U}^* f_{j(n)}) = 0$. But, from (iv)′, $(f_m, \hat{U}^* f_{j(m)}) = 1$, and hence $(f_m, \hat{U}^* f_{j(n)}) = 1$, since $j(m) = j(n)$, which is a contradiction. QED.

Hence, from (v),

(vi) $((f_{j(n)}, \hat{U} f_1) + (f_{j(n)}, \hat{U} f_2))/\sqrt{2} = 1$.

But, from (iv) and (iv)′,

(vii) $(f_{j(n)}, \hat{U} f_i) = 0$ for all $i \neq n$,[2] and

(vii)′ $(f_{j(n)}, \hat{U} f_n) = 1$.

Now there are three possibilities: $n = 1, 2$, or > 2. In each of these cases we see that substituting (vii) and (vii)′ into (vi) gives a contradiction, i.e. we get $(1 + 0)/\sqrt{2} = 1$, $(0 + 1)/\sqrt{2} = 0$, and $(0 + 0)/\sqrt{2} = 1$ respectively. Hence our initial supposition (i) is false. QED.

Note however that there is no bar to merely supposing that (i) holds for a particular choice of $\{c_i\}$. It is (i) holding for *all* choices of $\{c_i\}$, even though f_j is allowed to depend on which $\{c_i\}$ is chosen, which leads to the contradiction.

[2] Since $(f, \hat{U}g)^* = (\hat{U}g, f) = (g, \hat{U}^*f)$ and $(\hat{U}^*)^* = \hat{U}$ for unitary operators.

APPENDIX 9

The Born Interpretation

Consider the measurement of a non-degenerate physical quantity Q for system S at time t, with eigenvector f_i in $\mathcal{H}(S)$ for eigenvalue q_i for all i. The measurement is an interaction between S and a measuring apparatus M from time t to t'. We do not assume it conserves Q, but assume M is in an initial pure state F_0 at t. (The generalization to an initial mixed state is easy—see Krips (1969b).) Moreover we assume that if $f(S, t) = f_i$ then M at t' registers q_i in the following sense: the state of M at t' is a mixture (although it may be a trivial mixture, i.e. pure) of eigenvectors of a physical quantity $Q(M)$ for M for its ith eigenvalue. Thus for each i

$$\hat{U}_{t', t}(S + M)f_i \times F_0 = \Phi_i,$$

where Φ_i is a vector in $\mathcal{H}(S + M)$, and $\hat{W}(M, t')$, which is just $\text{Tr}_S \hat{P}(F_i)$,[1] must be a mixture of eigenvectors of $Q(M)$ for the ith eigenvalue. Note here that I am explicitly allowing $Q(M)$ to be degenerate. So far I have not needed to do this, because I was only considering conservative measurements which forced Φ_i to take the form $f_i \times F_i$.

Now let F_{id} for fixed i and varying d be some maximal set of orthogonal eigenvectors of $Q(M)$ for the ith eigenvalue, so that $(F_{id}, F_{je}) = \delta_{ij}\delta_{de}$. Then quite generally (since $\{F_{id}\}$ is complete in $\mathcal{H}(M)$ and $\{f_i\}$ is complete in $\mathcal{H}(S)$)

$$\Phi_i = \sum_{k, j, e} c^i_{kje} f_k \times F_{je}$$

for some $\{c^i_{kje}\}$. The condition that $\text{Tr}_S \hat{P}(\Phi_i)$ is a mixture of eigenvectors of $Q(M)$ for the ith eigenvalue then implies that with no loss of generality we can write, for some $\{c^i_{ke}\}$,

(i) $\quad \Phi_i = \sum_{ke} c^i_{ke} f_k \times F_{ke}.$

(We do this by showing that $c^i_{kje} = 0$ for $i \neq j$ and letting $c^i_{kie} = c^i_{ke}$.) And, since $(\Phi_j, \Phi_j) = 1$,

(i)' $\quad \sum_{k, e} c^j_{ke} \bar{c}^j_{ke} = 1.$

---

[1] By the principle (von Neumann) of Appendix 6.

Moreover, since $\hat{U}_{t',t}$ (abbreviating $\hat{U}_{t',t}(S+M)$) is linear, we have that

$$\hat{U}_{t',t}\,(\Sigma\,c_i f_i)\times F_0 = \Sigma\,c_i\Phi_i.$$

We now require that for some maximal set $\{F_{id}\}$ of orthogonal eigenvectors of $Q(M)$ and some $\{p_{id}\}$ for which $\Sigma_{i,d}\,p_{id}=1$ it is the case that

(ii) $\text{Tr}_S\,\hat{P}(\Sigma\,c_i\,\Phi_i) = \sum_{i,d} p_{id}\,\hat{P}(F_{id}).$

This last equality expresses the requirement that if the initial state of S is $\Sigma\,c_i f_i$, whatever the $\{c_i\}$, then the measurement of Q on S determinately registers some value for Q, i.e. $Q(M)$ determinately has its ith value for some i. (At work here is the criterion (Bohr)' for $Q(M)$ to have a determinate value.)

Now take $(F_{je},\ldots,F_{kf})$ of both sides of (ii). The LHS

$$= \left(F_{je}\times\sum_l \left(f_l,\sum_{m,n}c_m\bar{c}_n\Phi_m\right)(\Phi_n,f_l\times F_{kf})\right)$$

$$= \sum_{m,n,l}c_m\bar{c}_n\,(f_l\times F_{jl},\Phi_m)\,(\Phi_n,f_l\times F_{kf}).$$

But $(F_{je},\Phi_m)=0$ for $m\neq j$ (by (i)'). Hence LHS

$$= \sum_l c_j\bar{c}_k\,(f_l\times F_{je},\,\Phi_j)\,(\Phi_k,f_l\times F_{kf})$$

$$= c_j\bar{c}_k\sum_l c_{le}^j\,\bar{c}_{ef}^k.$$

But the RHS

$$= p_{je}\delta_{jk}\delta_{ef}.$$

Hence

$$p_{je}\delta_{jk}\delta_{ef} = c_j\bar{c}_k\sum_l c_{le}^j\bar{c}_{ef}^k.$$

Now set $e=f$, $j=k$ and sum over e. This gives:

$$\sum_e p_{je} = |c_j|^2\sum_{l,e}c_{le}^j\,\bar{c}_{le}^j,$$

which, by (i)', means that

(iii) $\sum_d p_{id} = |c_i|^2.$

Now consider a non-degenerate physical quantity $Q'(M)$ for M with the $\{F_{id}\}$ as a complete set of eigenvectors, and for which, for any i, $Q(M)$ has its ith value iff $Q'(M)$ has one of the values q_{id} for some d, and with the same probability. (It is an assumption that any degenerate physical quantity is related to some non-degenerate physical quantity in this way—as a function of it. The consequences of this are discussed in Chapter 7.) Then, by (Bohr)′, and since

$$\hat{W}(M, t') = \sum p_{id}\, \hat{P}(F_{id}),$$

it follows that $\Sigma_d p_{id}$ is the probability that $Q'(M)$ has one of the values $q_{i1}, q_{i2}, \ldots$ *Ex hypothesi* this is also the probability that $Q(M)$ has its ith value at t, which is also the probability that Q registers the value q_i at t', in conformity with (BSI). The same result can also be obtained more directly from (Bohr)″ of Appendix 5. Thus (BSI)** is derived for this more general class of measurements. QED.

The Principal Principle

Proof 1

Assume $P(E$ has $R) < 1$. Hence either (a) or (b) of the main text holds. In what follows we shall let '$\{S_n\}$', 'm' respectively denote the sets of spheres of possible worlds and the continuous measure which satisfy (a) and (b), whichever it is that holds. Now assume (as premiss in a *reductio*)

(P) There is an S^* for which $m(E.S^*) \neq 0$, and all but a measure 0 of the E-worlds in S^* are R-worlds.

Hence $m(E.S^*) = m(E.S^*.R)$, so that $m(E.R.S^*)/m(E.S^*.) = 1$. And it follows that

(P)′ $m(E.R.S)/m(E.S) = 1$ for any $S < S^*$ for which $m(E.S) \neq 0$ (since for any $S \leq S^*$, all E-worlds must also be R-worlds).

Now consider the following two cases.

The first case

$\{S_n\}$ and m satisfy (a). Then, by $(P)′$, for any S_n for which $S_n \geq S^*$ and $m(E.S_n) \neq 0$, it is the case that $m(E.R.S_n)/m(E.S_n) = 1$. Hence, since for all S_n here $m(E.S_n) \neq 0$, it follows that $P(E$ has $R) = \lim_{n \to \infty} 1 = 1$, which contradicts the assumption that $P(E$ has $R) < 1$.

The second case

$\{S_n\}$ and m satisfy (b). Now we can quickly prove, by *reductio*, that S^* is an upper bound on the S_n for which $m(E.S_n) = 0$. Suppose, as premiss of the *reductio*, that there is some S_m for which $S_m > S^*$ and $m(E.S_m) = 0$. But by Munroe (1955) theorem 27.1, p. 191, $m(E.S^*) \leq m(E.S)$ for any S for which $S^* < S$. Hence $m(E.S^*) \leq m(E.S_m)$, and so $m(E.S^*) = 0$. But $m(E.S^*) > 0$, by (P), and so we have our contradiction. Therefore, as required, S^* is an upper bound. Hence, by definition of $\bar{S}$ as a *least* upper bound, $\bar{S} \leq S^*$. Hence, by $(P)′$ above, $m(E.R.\bar{S})/m(E.\bar{S}) = 1$. Hence, since we assume the continuity of the measure m,

$$\lim_{S_n \to \bar{S}} m(E.R.S_n) = 1.$$

And so, by (b), $P(E$ has $R) = 1$, which contradicts the assumption that $P(E$ has $R) < 1$.

Thus in either case we have a contradiction, and the *reductio* is completed, i.e. we have shown that if $P(E$ has $R) < 1$ then (P) must fail, which is just what the theorem in the main text says.

Proof 2

The proof of the Principal Principle will be carried out for its time-independent form:

$$C(A/X) = x,$$

where X is the proposition that the objective chance of A is x, and C is any reasonable credence function.[1] The 'objective chance of A' I define along exactly the lines of the possible world analysis for objective single-case probabilities given in the text, i.e. as either

$$P(A) = \text{limit}_{S_n \to \bar{S}} m(A.S_n)/m(S_n),$$

or

$$P(A) = \text{limit}_{n \to \infty} m(A.S_n)/m(S_n),$$

depending on whether the relevant $\bar{S}$ exists. I also assume that the credence function C is given by:

$$C(P) = m(P)$$

for that case where the value of C on the whole space of possible worlds has been normalized to 1. Note that implicit here is the non-trivial assumption that the measure over possible worlds, which is used to construct an a priori probability, viz. the credence function, is the same as the measure used to count possible worlds in setting up the a posteriori objective single-case probability.

To prove the Principal Principle I first of all need to generalize the notion of $P(A)$, the objective chance of A at the actual world, to $P_w(A)$, the objective chance of A at a general world w. This is easily done by introducing a notion, from measure theory, of a monotone sequence of nets for the measure m.[2] This is a sequence of classes of disjoint measurable sets (measurable according to m), each set of finite measure m, the sets in any one class being exhaustive, and every set in the $n + 1$th class being a subset of some set in the nth class, for all n. We then take it that

> For some measure m, and some monotone sequence of nets for m, $P_w(A)$ = $\lim_{n \to \infty} m(S_{nw}.A)/m(S_{nw})$, where S_{nw} is the set which contains w in the nth class of sets belonging to the relevant monotone sequence of nets.

---

[1] Lewis (1980) p. 231.   [2] See Munroe (1953), sect. 43.

(We will here only consider the case for which $m(S_{nw} . A) \neq 0$ for all S_{nw}. The other case is dealt with easily, along exactly the same lines followed in the definition of $P(A)$.)

In Munroe's notation this says:

$P_w(A) = D_{\mathcal{N}_m} m(w.A)$, for some monotone sequence of nets $\mathcal{N}$ for some measure m.

In addition it will be assumed that $\mathcal{N}$ is regular[3] with respect to m, which means:

For each m-measurable set E, there is a sequence K of subsets of the members of $\mathcal{N}$ such that for any $\varepsilon > 0$, $m(E - \cup K_i) = 0$ and $m(\cup K_i) \leq m(E) + \varepsilon$.

Now we can quickly complete the proof by using a standard theorem of measure theory, viz. Theorem 43.4 of Munroe (1953):

If $\mathcal{N}$ is a monotone sequence of nets, regular with respect to m, and m' is an everywhere finite, completely additive measure absolutely continuous[4] with respect to m, then $D_{\mathcal{N}_m} m'$ exists almost everywhere, and

$m' = \int D_{\mathcal{N}_m} m' dm$.

Define m^* by '$m^*(E) = m(E.A)$'. Obviously m^* is finite (since m is), and is completely additive (since m is), and absolutely continuous with respect to m (since $m(E.A) = 0$ if $m(E)$ is—see Theorem 27.1 of Munroe (1955) p. 191). Moreover, by our definition,

$P_w(A) = D_{\mathcal{N}_m} m^*(w)$.

Since $\mathcal{N}$ is regular with respect to m^*, by the above theorem it follows that

$m^* = \int P_w(A) \, dm$.

In particular, since $P_w(A) = x$ for any w in X, $m^* = \int x dm = x.m$ and hence

$m^*(X) = m(X.A) = x.m(X)$,

and hence

$m(X.A)/m(X) = x$.

And this means that $c(A/X) = P_w(A)$. QED.

The extension to a time-dependent version of the Principal Principle cannot be discussed here, but is straightforward.

---

[3] See Munroe (1953), p. 297.
[4] m' is absolutely continuous with respect to m iff $m'(E) = 0$ if $m(E) = 0$.

APPENDIX 11

The Independence of Pure States

We need to show that

$$P([(\mathbf{a}, \mathbf{b})_n \text{ and } (\mathbf{a}, \mathbf{b})_m], [(a, b) \text{ and } (a', b')], [L_n + R_n + L_m + R_m], t)$$
$$= P((\mathbf{a}, \mathbf{b})_n, (a, b), L_n + R_n, t) \times P((\mathbf{a}, \mathbf{b})_m, (a', b'), L_m + R_m, t).$$

This is shown by applying the Born interpretation to the system $L_n + R_n + L_m + R_m$. The left-hand side in the previous equality is then given by

$$\text{Tr}[\hat{W}(L_n + R_n + L_m + R_m, t)\,\hat{P}_{L,n}(\mathbf{a}, a)\,\hat{P}_{R,n}(\mathbf{b}, b)\,\hat{P}_{L,m}(\mathbf{a}, a')\,\hat{P}_{R,m}(\mathbf{b}, b')],$$

where $\hat{P}_{L,n}(\mathbf{a}, a)$ is the projection operator onto the eigenspace of $\mathbf{a}$ for value a in $\mathscr{H}(L_n)$, etc. But

$$\hat{W}(L_n + R_n + L_m + R_m, t) = \hat{P}_n(F) \times \hat{P}_m(F),$$

where $\hat{P}_n(F)$ is the projection onto F for $\mathscr{H}(L_n + R_n)$ and $\hat{P}_m(F)$ similarly. Hence the left-hand side is factorizable as

$$\text{Tr}[\hat{P}_n(F)\,\hat{P}_{L,n}(\mathbf{a}, a)\,\hat{P}_{R,n}(\mathbf{b}, b)]\,\text{Tr}[\hat{P}_m(F)\,\hat{P}_{L,m}(\mathbf{a}, a)\,\hat{P}_{R,m}(\mathbf{b}, b)],$$

which is just the right-hand side, as required.

APPENDIX 12

Contextualization

If we use (VRR) instead of (VR) then we modify val $Q_{|R}$ in (CVR) to val $Q_{|R}(R)$, and the val $Q \times I_{|O(A,\,A')}$ in (ii)' to val $Q \times I_{|O(A,\,A')}(O(A,A'))$. As indicated, we can write the latter simply as: val $Q \times I_{|O(A,\,A')}(A,A')$ (the final (A,A') signifying that a preparation for a joint measurement of A and A' has taken place).

We also generalize OLOC to:

OLOC*   val $Q \times I_{|O(A,\,A')}(C,D) = $ val $Q \times I_{|O(A,\,B')}(C,D)$

and put forward:

ELOC   val $Q \times I_{|O(A,\,A')}(C,D) = $ val $Q \times I_{|O(A,\,A')}(C,E)$.

The proof then proceeds as in section 2, but (iv) now becomes:

(iv)′ If val $A \times I_{|O(A,\,A')}(A,A') = a_i$ then
     val $I \times A'_{|O(A,\,A')}(A,A') = a'_i$,

and (v) becomes:

(v)′ If val $I \times A'_{|O(B,\,A')}(B,A') = a'_i$ then
     val $Q \times I_{|O(B,\,A')}(B,A') = h(a_i)$.

But ELOC and OLOC* entail:

(vi)′ val $I \times A_{|O(A,A')}(A,A') = $ val $I \times A'_{|O(B,\,A')}(B,A')$.

Hence from (iv)′, (v)′, (vi)′ we derive:

(vii)′ If val $A \times I_{|O(A,\,A')}(A,A') = a_i$ then
     val $Q \times I_{|O(B,\,A')}(B,A') = h(a_i)$ if $\hat{Q} = h(\hat{A})$,

and the contradiction then follows as in Chapter 9, section 2.

Index of Principles

(Ind)   Whether or not S_2 is open has no causal relevance to the trajectories of electrons which go through S_1.

(Inf)   Experiment E has result R
$\therefore$ The probability of E having result R is 1.

THE BORN STATISTICAL INTERPRETATION

(BSI)   For non-degenerate discrete valued Q, with eigenvectors $\{f_i\}$, and for S at t in the pure state f,

$$P(Q, i, S, t) = |(f_i, f)|^2.$$

(BSI)'   For degenerate discrete valued Q, with $\{f_{i1}, f_{i2}, \ldots\}$ as a basis set for the subspace of eigenvectors for possible value q_i, and for S at t in pure state f,

$$P(Q, i, S, t) = \sum_x |(f_{ix}, f)|^2.$$

(BSI)*   For degenerate discrete valued Q, with S at t in the mixed state $\{p_n, f_n\}$,

$$P(Q, i, S, t) = \sum_n p_n \sum_x |(f_{ix}, f_n)|^2.$$

(BSI)** $P(Q, i, S, t) = \operatorname{Tr} \hat{W} \hat{P}_i$, where $\hat{P}_i$ is the projection operator onto the space of eigenvectors of Q for possible value q_i and $\hat{W}$ is the density operator of S at t.

DETERMINATE-VALUE PRINCIPLES

(Princ)   If $P(Q, i, S, t) = P(Q, i, S, t')$ for all Q, i then the same physical quantities have determinate values in S at t with the same probabilities as in S at t'.

(Princ)'   If there is probability p_n that the non-degenerate Q has value q_n determinately in S at t, and $\Sigma p_n = 1$, then S at t is in the mixed state $\{p_n, f_n\}$ (where f_n is the eigenvector of Q for possible value q_n, for all n).

(Princ)″   Even for degenerate Q if S at t is in the mixed state $\{p_n, f_n\}$ where f_n is an eigenvector of Q for possible value q_n for all n, and $q_n \neq q_m$ for any $n \neq m$, then there is probability p_n that Q has value q_n determinately in S at t for all n.

(Defn)   For any S, t, the density operator $\hat{W}(S, t)$ for S at t is $\Sigma\, p_n \hat{P}(f_n)$ if S at t is in the mixed state $\{p_n, f_n\}$.

(Bohr) Q has determinate value q in S at t if $f(S, t)$ is an eigenvector of Q.

(Bohr)′   If $\hat{W}(S, t) = \Sigma p_n \hat{P}(f_n)$, where f_n is an eigenvector of Q for possible value q_n and $q_n \neq q_m$ for all $n \neq m$ then there is probability p_n that Q has the value q_n determinately in S at t for all n.

INVALID CONVERSES TO (PRINC)″

(Weak   If there is probability p_n that Q, a non-degenerate q-quantity
Converse)   for S, has value q_n determinately for all n, then S at t is in a mixture of eigenvectors of Q for which the sum of the probabilities associated with eigenvectors for possible values q_n is p_n for all n.

(Weaker   If Q has a determinate value in S at t then S at t is in a (possibly
Converse)   trivial) mixture of Q eigenvectors.

MEASUREMENT

(Cal)   An ideal measurement of Q in S at t is any interaction between S and a macroscopic system M, which starts at t and terminates at t', for which there is a physical quantity $Q(M)$ such that for all i were $f(S, t)$ an eigenvector of Q for possible value q_i then $Q(M)$ would take its ith value with certainty, where the various values of $Q(M)$ are macroscopically distinct; and

(Cal)′ Q is measured to have value q_i in S at t iff $Q(M)$ takes its ith value at t'.

THE REDUCTION OF THE WAVE-PACKET (FOR CONSERVATIVE MEASUREMENTS, IN WHICH M IS INITIALLY IN A PURE STATE)

Pure-pure   If $f(S, t) = \Sigma c_i f_i$ then, after measurement of S at t, S is in the
form   pure state f_i for some i.

Pure-mixed form	If $f(S, t) = \Sigma c_i f_i$ then, after measurement of S at t, $S + M$ is in a mixture which has a density operator of the form:

$$\sum |c_i|^2 \hat{P}(f_i) \times \hat{P}(F_i)$$

($\{F_i\}$ being macroscopically distinct states of M).

THE PROJECTION POSTULATE

If $f(S, t) = \sum c_i f_i$ then after measurement S is in a mixed state with density operator

$$\sum |c_i|^2 \hat{P}(f_i).$$

(Von Neumann) $\hat{W}(S_1, t) = \mathrm{Tr}_2 \, \hat{W}(S_1 + S_2, t)$

REALIST PRINCIPLES FOR QT

(Det Q)	Physical quantities always have determinate values
(Pass Q)	The measured value of Q in S at t = the value possessed by Q in S at t.
(NDQ)	The value possessed by the measured quantity just after measurement = the value registered by the measurement, i.e. the measured value.

Note 1. The measurements referred to in (NDQ) and (Pass Q) are implicitly taken as the *ideal* measurements of QT, and the physical quantities are implicitly taken to be q-quantities.

Note 2. The above principles are sometimes presented in generalized form, as applying to ideal measurements and physical quantities at large. In that case we write them as (Det), (Pass), and (ND).

GLEASON'S THEOREM COROLLARY

(G)'	There is no single-valued map m from the set of normalised vectors in an at least three dimensional Hilbert space onto 1 or 0 for which $\Sigma m(f_i) = 1$ for any complete mutually orthogonal set of normalized vectors $\{f_i\}$.

COMPLETENESS

(U)   If Q,Q' have the same single eigenvector in $\mathscr{H}(S)$ for eigenvalues q,q' respectively then $Q(S,t)=q$ iff $Q'(S,t)=q'$.

$(U)_0$ If Q and Q' have all their eigenvectors in common as well as the corresponding possible values, then a Q measurement is a Q' measurement; and . . .

$(U)'_0$ Q is measured to have value q'_i in S at t iff Q' is measured (by the same measurement) to have value q'_i in S at t.

$(U)_1$ If $\hat{Q}=\hat{Q}'$ then $Q=Q'$.
(special case
of (U))

$(U)'$ If Q and Q' have the same single eigenvector in $\mathscr{H}(S)$ for eigenvalues q,q' respectively then $mv(Q,S,t)=q$ iff $mv(Q',S,t)=q'$.

$(U)^*$ For non-degenerate Q,Q', if $\hat{Q}=\hat{Q}'$ then $Q=Q'$

(C)   For any complete orthonormal set of vector $\{f_i\}$ in $\mathscr{H}(S)$ there is a non-degenerate Q for S for which $\{f_i\}$ are its eigenvectors.

(C1)   For any non-degenerate q-quantity Q for S and any possible value q_i there is a two-valued q-quantity Q_i for S for which

$Q_i(S,t)=1$ if $Q(S,t)=q_i$.
$Q_i(S,t)=0$ if $Q(S,t)=q_j$ for $j\neq i$
$P(\text{val } Q_i,1,S,t)=1$ if $P(\text{val } Q,i,S,t)=1$
$P(\text{val } Q_i,0,S,t)=1$ if $P(\text{val } Q,j,S,t)=1$
for any $j\neq i$.

(C2)   For any Q for which $\hat{Q}=\hat{Q}_i$, $Q_i(S,t)=Q(S,t)$ for all t.

FUNCTIONS OF QUANTITIES

(Func)   If $\hat{Q}'=F(\hat{Q})$ then $Q'=F(Q)$.
$(Func)_1$ If Q is a q-quantity then so is $F(Q)$ and $\hat{F}(Q)=F(\hat{Q})$.
$(Func)_2$ If $\hat{Q}'=F(\hat{Q})$ then $Q'=F(Q)$.
$(Func)_0$ $mv(F(Q),S,t)=F(mv(Q,S,t))$ for non-degenerate Q.
(Func 1)   If $\hat{Q}'=F(\hat{Q})$ for $\hat{Q}$ a one-dimensional projector then $Q'=F(Q)$.
$(Func)^*$ If $\hat{Q}=f(\hat{A})$ for $\hat{A}$ maximal then there is some q-quantity $Q_{|A}$, represented by $\hat{Q}$, for which $Q_{|A}=f(A)$.
(Func R)   If $\hat{Q}=h(\hat{A})$ and val $A\times I_{|(A,A')}=a_i$ then val $Q\times I_{|O(B,A')}=h(a_i)$ for B,A non-degenerate on S_1 and A' non-degenerate on S_2 and $\hat{Q}$ a function of $\hat{B}$.

MEASURED VALUES

(MV)   For every Q, S, t there is a (counterfactual) measured value $mv(Q, S, t)$, the value which Q would be measured to have were it measured in S at t.

PASSIVITY PRINCIPLES

$(\text{Pass } Q)_1$ For any (ideal) method M of measuring Q in S at t if $mv_M(Q, S, t)$ and $Q(S, t)$ exist then $mv_M(Q, S, t) = Q(S, t)$.

$(\text{U—Pass } Q)_1$ If $mv(Q, S, t)$ and $Q(S, t)$ exist then $mv(Q, S, t) = Q(S, t)$.

$(\text{R—Pass } Q)_1$ For some (ideal) method M if $mv_M(Q, S, t)$ and $Q(S, t)$ exists then $mv_M(Q, S, t) = Q(S, t)$.

$(\text{Pass } Q)_2$ For any M, if $Q(S, t)$ exists then so does $mv_M(Q, S, t)$ and $mv(Q, S, t) = Q(S, t)$.

$(\text{U—Pass } Q)_2$ If $Q(S, t)$ exists then so does $mv(Q, S, t)$ and $mv(Q, S, t) = Q(S, t)$.

$(\text{R—Pass } Q)_2$ For some M if $Q(S, t)$ exists then so does $mv_M(Q, S, t)$ and $mv_M(Q, S, t) = Q(S, t)$.

$(\text{Pass } Q)'_2$ For any M if $mv_M(Q, S, t)$ exists then so does $Q(S, t)$ and $mv_M(Q, S, t) = Q(S, t)$.

$(\text{U—Pass } Q)'_2$ If $mv(Q, S, t)$ exists then so does $Q(S, t)$ and $mv(Q, S, t) = Q(S, t)$.

$(\text{R—Pass } Q)'_2$ For some M if $mv_M(Q, S, t)$ exists then so does $Q(S, t)$ and $mv_M(Q, S, t) = Q(S, t)$.

$(\text{Pass } Q)'$ Q has the value q in S at t with probability 1 iff $P(Q, q, S, t) = 1$.

(Loc Pass)   Were Q measured on S at t then it would be the case that the measured value equals the value possessed by Q in S at t.

$(\text{Pass } Q)^*$ $mv(A, S, t) = A(S, t)$ for maximal A.

FREEDOM OF MEASUREMENT (IN THESE PRINCIPLES C IS A SET OF N ELECTRON PAIRS)

$(\text{FM})_1$ Were $(\mathbf{a}, \mathbf{b})$ to be measured on each of the members of C at t, the laws of QT would still hold.

$(\text{FM})_2$ Were $(\mathbf{a}, \mathbf{b})$ to be measured on each of the members of C at t, each of the members of C would still be in the same state at t.

LOCALITY

(Strong Loc)   Were $(\mathbf{a}, \mathbf{b})$ measured on each of the members of C at t then the mv of $\mathbf{a}$ for L_n at t would still be a_n (where L_n is the left-hand member of the nth pair of electrons, and 'a_n' is a rigid designator for the value which **a** would be measured to have were it measured on L_n at t irrespective of what else is measured on the members of C at t).

HEYWOOD–REDHEAD

(OLOC)   If S_1 and S_2 are spatially separated and A is non-degenerate on S_1 then for any X, Y non-degenerate on $S_1 + S_2$

$$\text{val } A \times I_{|X} = \text{val } A \times I_{|Y}.$$

(AWOLOC)   If S_1 and S_2 are spatially separated then for A', A' non-degenerate on S_2 and A non-degenerate on S_1, and $\hat{Q}$ a function of $\hat{A}$,

$$\text{val } Q \times I_{|O(A, A')} = \text{val } Q \times I_{|O(A, B')}.$$

(BWOLOC)   As for (AWOLOC) but S_1, S_2 interchanged.

(ELOC) $\text{val } Q \times I_{|O(A, A')}(C, D) = \text{val } Q \times I_{|O(A, A')}(C, E)$, for S_1, S_2 spatially separated.

(VR)   If $P(Q, i, S, t) = 0$ then it is not the case that $Q(S, t) = q_i$.

(ER)   If $f(S, t) = f_i$, where f_i is eigenvector of Q for possible value q_i, then $Q(S, t)$ exists $= q_i$.

(ER)*   If $f(S, t) = f_i$ then $P(Q(S, t), q_i) = 1$.

(ER)'   If S at t is in $\{p_n, f_n\}$ where each f_n is an eigenvector of Q for possible value q_i then $Q(S, t)$ exists $= q_i$.

(VR)'   If $\text{Prob } (Q, i, S, t) = 1$ then $Q(S, t) = q_i$.

(GVR)   If $\text{Prob } (Q, I, S, t) = 0$ then it is not the case that $Q(S, t) \in I$ (where I is an interval of values for Q).

(GVR)'   If $\text{Prob } (Q, I, S, t) = 1$ then $Q(S, t) \in I$.

(VRR)   If $\text{Prob } (Q, i, S, t) = 0$ then it is not the case that $Q(S, t)(Q) = q_i$ (where $Q(S, t)(Q)$ is the value possessed by Q in S at t when S at t has been prepared for a Q-measurement).

(CVR)   If R is maximal on S, and there are functions f, f' for which $\hat{Q} = f(\hat{R})$ and $\hat{Q}' = f'(\hat{R})$, and if $\text{Prob } ((Q, Q'), (q, q')) = 0$ then either $\text{val } Q_{|R} \neq q$ or $\text{val } Q'_{|R} \neq q'$.

(ERR)   If $f(S, t)$ is an eigenvector of Q for possible value q then $Q(S, t)(Q) = q$.

$(P)_0$ A is represented by $\hat{A}$ on $\mathscr{H}(S_1)$ iff it is also represented by $\hat{A} \times \hat{I}$ on $\mathscr{H}(S_1 + S_2)$.

$(P)_1$ If A is a q-quantity for both S and S' then $A(S, t) = A(S', t)$.

Bibliography

Araki, H. and Yanase, M. (1960). *Physical Review* **120**, 622.

Aspect, A., Dalibard, J., and Roger, G. (1982). *Phys. Rev. Letters* **49**, 1804.

Belinfante, F. (1973). *A Survey of Hidden-variable Theories* (Pergamon Press, Oxford).

Bell, J. (1966) *Reviews of Modern Physics* **38**, 448.

Bell, J. (1981). *Journal de Physique* **42** (C2), 41.

Bohm, D. and Aharanov, Y. (1960). *Il Nuovo Cimento* **17**, 964.

Bohr, N. (1928). *Nature* **121**, 580.

Bohr, N. (1934). *Atomic Theory and the Description of Nature* (Cambridge University Press, Cambridge).

—— (1935). *Physical Review* **48**, 696.

—— (1949). In *Albert Einstein Philosopher–Scientist*, ed. P. Schilpp (Evanston University Press, Evanston, Ill.).

Born, M. (1926). *Zeitschr. f. Phys.* **37**, 863.

—— (1926). *Zeitschr. f. Phys.* **38**, 803.

—— (1949). *Natural Philosophy of Cause and Chance* (Clarendon Press, Oxford).

Carnap, R. (1966). *Philosophical Foundations of Physics*, ed. M. Gardner (Basic Books, New York, NY).

Cartwright, N. (1974). *Synthese* **29**, 229.

—— (1983). *How the Laws of Physics Lie* (Cambridge University Press, Cambridge).

Clauser, J., Holt, R., Horne, M., and Shimony, A. (1969). *Phys. Rev. Letters* **23**, 830.

—— and Shimony, A. (1978). *Reports on Progress in Physics* **41**, 1881.

De Broglie, L. (1923). *C. R. Acad. Sci. Paris* **177**, 507.

Daneri, A., Loinger, A., and Prosperi, G. (1962). *Nuclear Physics* **33**, 197.

D'Espagnat, B. (1979). *Scientific American* **241**, 128.

—— (1971). *Conceptual Foundations of Quantum Mechanics* (Benjamin, New York, NY).

Dirac, P. (1958). *The Principles of Quantum Mechanics* (Clarendon Press, Oxford).

Dummett, M. (1975). *Synthese* **30**, 325.

—— (1979). In *Perception and Identity*, ed. G. McDonald (Macmillan, London).

Eberhard, P. (1977). *Il Nuovo Cimento* **38B**, 75.

Einstein, A., Podolski, B., and Rosen, N. (1935). *Physical Review* **47**, 777.

Everett, H. (1957). *Reviews of Modern Physics* **29**, 454.

Feller, W. (1957). *An Introduction to Probability Theory and Its Applications* (Wiley, New York, NY).

Feyerabend, P. (1962). In *Frontiers of Science and Philosophy*, ed. R. Colodny (Pittsburgh University Press, Pittsburgh, Pa.).

Feynman, A., Leighton R., and Sands, M. (1965). *The Feynman Lectures on Physics* **3** (Addison–Wesley, Reading, Mass.).

Fine, A. (1971). In *Problems in the Foundations of Physics* **4**, ed. M. Bunge (Springer-Verlag, Berlin).

—— (1979). *Synthese* **42**, 145.

Fine, K. (1975). *Synthese* **30**, 265.

Gleason, A. (1957). *Journal of Mathematics and Mechanics* **6**, 885.

Gould, W. (1983). *Mammals of Australia*, ed. J. Dixon (Macmillan, Melbourne, Australia).

Grangier, P., Roger, G., and Aspect, A. (1986). *Europhysics Letters* **1**, 173.

Haack, S. (1974). *Deviant Logic* (Cambridge University Press, Cambridge).

Hacking, I. (1971). *Representing and Intervening* (Cambridge University Press, Cambridge).

Hartle, J. (1968). *American Journal of Physics* **36**, 704.

Healey, R. (1979). *Synthese* **42**, 121.

Heisenberg, W. (1930). *The Physical Principles of the Quantum Theory* (Dover, New York, NY).

—— (1971). *Physics and Beyond* (Allen & Unwin, London).

Hellman, G. (1982a). *Synthese* **53**, 445.

—— (1982b). *Synthese* **53**, 461.

Hesse, M. (1961). *Forces and Fields* (Nelson, London).

Heywood, P. and Redhead, M. (1983). *Foundations of Physics* **13**, 481.

Jammer, M. (1974). *The Philosophy of Quantum Mechanics* (Wiley, New York, NY).

Jauch, J. (1968). *Foundations of Quantum Mechanics* (Addison–Wesley, New York, NY).

Kochen, S. and Specker, E. (1967). *Journal of Mathematics and Mechanics* **17**, 59.

Krips, H. (1969a). *Il Nuovo Cimento*, Ser. I, **6**, 1127.

—— (1969b). *Il Nuovo Cimento* **61B**, 12.

—— (1986). *Studies in History and Philosophy of Science* **17**, 43.

——(1987). *Realism and Collapse of the Wave-packet* (forthcoming).

Lakatos, I. (1970). *Criticism and the Growth of Knowledge*, ed. I. Lakatos and A. Musgrave (Cambridge University Press, Cambridge).

Landé, A. (1965). *New Foundations of Quantum Mechanics* (Cambridge University Press, Cambridge).

Lemmon, E. (1965). *Beginning Logic* (Nelson, Sunbury-on-Thames).

Lewis, D. (1973). *Counterfactuals* (Blackwell, Oxford).

—— (1975). In *Causation and Conditionals*, ed. E. Sosa (Oxford University Press, Oxford).

——(1980). In *Inductive Logic II*, ed. R. Jeffrey (Unversity of California Press, Berkeley, Calif.).

London, F. and Bauer, E. (1983). In *Quantum Theory and Measurement*, ed. J. Wheeler and W. Zurek (Princeton University Press, Princeton, NJ).

Ludwig, G. (1954). *Die Grundlagen der Quantenmechanik* (Springer, Heidelberg).

Mandel, L. and Pfleegor, R. (1967). *Physical Review* **159**, 1084.

Mackie, J. (1975). In *Causation and Conditionals*, ed. E. Sosa (Oxford University Press, Oxford).

McKinnon, E. (1980). *Studies in the Foundations of Quantum Mechanics*, ed. P. Suppes (Philosophy of Science Association, East Lansing, Mich.).

Mellor, H. (1981). In *What, Where, When, Why*, ed. R. Mclaughlin (Reidel, Dordrecht).

—— (1971). *The Matter of Chance* (Cambridge University Press, Cambridge).

Munroe, M. (1953). *Measure and Integration* (Addison–Wesley, Reading, Mass.).

Pais, A. (1982). *Subtle is the Lord* (Oxford University Press, Oxford).

Pauli, W. (1933). *Handbuch der Physik* **24**, ed, H. Geiger and K. Scheel (Springer, Berlin).

Pipkin, F. (1978). *Adv. At. Mol. Phys.* **14**, 281.

Poincaré, H. (1957). *Science and Hypothesis* (Dover, New York, NY).

Popper K. (1982a). *The Open Universe* (Hutchinson, London).

—— (1982b). *Realism and the Aim of Science* (Hutchinson, London).

—— (1982c). *Quantum Theory and the Schism in Physics* (Hutchinson, London).

Putnam, H. (1969). In *Boston Studies in the Philosophy of Science* **5**, ed. R. Cohen and M. Wartofsky (Reidel, Dordrecht).

Reichenbach, H. (1944). *Philosophical Foundations of Quantum Mechanics* (University of California Press, Berkeley, Calif.).

—— (1956). *The Direction of Time* (University of California Press, Berkeley, Calif.).

Rorty, R. (1980). *Philosophy and the Mirror of Nature* (Blackwell, Oxford).

Salmon, W. (1975). In *Explanation*, ed. S. Körner (Yale University Press, New Haven, Conn.).

Schiff, L. (1955). *Quantum Mechanics* (McGraw-Hill, New York, NY).

Schrödinger, E. (1926a). *Phys. Zeitschr.* **27**, 95.

—— (1926b). *Ann. d. Phys.* **79**, 361.

—— (1935). *Naturwissenschaften* **48**, 552.

Scriven, M. (1962). *Minnesota Studies in the Philosophy of Science* **3**, ed. G. Maxwell and H. Feigl (Minnesota University Press, Minneapolis, Minn.).

Shimony, A. (1984). *B.J.P.S.* **35**, 25.

Stapp, H. (1971). *Physical Review* **D3**, 1303.

—— (1977). *Foundations of Physics* **7**, 313.

Van Frassen B. (1973). In *Contemporary Research in the Foundations and Philosophy of Quantum Theory* **2**, ed. C. Hooker (Reidel, Dordrecht).

——(1974). In *Boston Studies in the Philosophy of Science* **15**, ed. R. Cohen and M. Wartofsky (Reidel, Dordrecht).

——(1979). *Synthese* **42**.

——(1980). *The Scientific Image* (Clarendon Press, Oxford).

Von Neumann, J. (1929). *Math. Ann.* **36**, 92.

——(1955). *Mathematical Foundations of Quantum Mechanics* (Princeton University Press, Princeton, NJ).

Wessels, L. (1960). *Studies in the Foundations of Quantum Mechanics*, ed. P. Suppes (Philosophy of Science Association, East Lansing, Mich.).

Wigner, E. (1952). *Zeitschr. f. Phys.* **133**, 101.

——(1967). *Symmetries and Reflections* (Indiana University Press, Bloomington, Ind.).

Zukav, G. (1980). *The Dancing Wu-Li Masters* (Fontana, London).

Index